*Analysis of*
*Energy Efficiency of*
*Industrial Processes*

Vladimir S. Stepanov

# Analysis of Energy Efficiency of Industrial Processes

With 38 Figures and 31 Tables

Springer-Verlag
Berlin Heidelberg New York
London Paris Tokyo
Hong Kong Barcelona
Budapest

Professor Dr. Vladimir S. Stepanov
Sibiran Energy Institute, Lermontov Street 130,
664033 Irkutsk, Russia

ISBN-13: 978-3-642-77150-7 e-ISBN-13: 978-3-642-77148-4
DOI: 10.1007/978-3-642-77148-4

Library of Congress Cataloging-in-Publication Data.
Stepanov, V. S. (Vladimir Sergeevich) [Analiz energeticheskogo sovershenstva technologicheskikh protsessov. English] Analysis of energy efficiency of industrial processes / Vladimir S. Stepanov. p. cm. Translation of: Analiz energeticheskogo sovershenstva tekhnologicheskikh protsessov. Includes bibliographical references and index. 1. Industry–Energy conservation. 2. Metallurgical plants–Energy conservation. I. Title. TJ163.3.S7413 1992 670.42–dc20 92-33780


Softcover reprint of the hardcover 1st edition 1993



Typesetting: Camera ready by the authors
57/3140 - 5 4 3 2 1 0 - Printed on acid-free paper

# Preface

It is universally recognized that the end of the current and the beginning of the next century will be characterized by a radical change in the existing trends in the economic development of all countries and a transition to new principles of economic management on the basis of a resource and energy conservation policy. Thus there is an urgent necessity to study methods, technical aids and economic consequences of this change, and particularly, to determine the possible amounts of energy resources which could be conserved (energy "reserves") in different spheres of the national economy.

An increased interest towards energy conservation in industry, one of the largest energy consumers, is quite natural and is manifested by the large number of publications on this topic. But the majority of publications are devoted to the solution of narrowly defined problems, determination of energy reserves in specific processes and plants, efficiency estimation of individual energy conservation measures, etc. However, it is necessary to develop a general methodological approach to the solution of such problems and create a scientific and methodical base for realizing an energy conservation policy.

Such an effort is made in this book, which is concerned with methods for studying energy use efficiency in technological processes and estimation of the theoretical and actual energy reserves in a given process, technology, or industrial sector on the basis of their complete energy balances.

The book generalizes the results of studies in this field conducted over many years at plants of ferrous and non-ferrous metallurgy with the use of exergy analysis, which were performed by the author himself and with co-authors. These works were started in the 1960s at the Kazakh Energy Institute under the directory of I. Kh. Ozoling and are continued now in the Siberian Energy Institute, Siberian Branch of the Russian Academy of Sciences.

This book is of interest to diverse audience of specialists in the areas of thermodynamics, industrial energetics, and energy economy.

Data for the calculation of energy balances of processes and plants in Chaps. 5, 6 are obtained in a study conducted by the author at plants operating in the USSR. The calculations described in Chap. 8 are based on official statistics, on the data of reports and forecasts from the USSR Ministry of Metallurgy.

Chapters 7 and 8 are written jointly with T. B. Stepanova whose help at all the stages of work on the monograph is gratefully acknowledged. I wish to express my gratitude to V. P. Ermakova, V. G. Borovikova and V. I. Viryukina for translating this book into English.

Irkutsk, March 1992 *Vladimir S. Stepanov*

# Contents

# Nomenclature

| | |
|---|---|
| $A$ | work, J |
| $\Delta A$ | work losses, J |
| $A$ | ash content in fuel |
| $a$ | fuel cost, currency /kg c.e. (coal equivalent) |
| $B$, $b$ | anergy (total and specific), J |
| $C$ | specific heat capacity, J/kg · K |
| $C$, $c$ | annual costs, currency/year; currency/year · unit of process product |
| $D$ | damage, monetary |
| $d$ | number of considered technological schemes |
| $E$, $e$ | exergy (total and specific), J, kW h; J/kg, J/mol, kW h/ton |
| $EF$ | economic effect, currency/year |
| $F$ | Helmholtz free energy, J |
| $F$ | Faraday constant, k Coulomb/mol |
| $G$ | Gibbs free energy, J |
| $\Delta G^0_z$ | standard Gibbs energy of formation of substance $z$, J |
| $\Delta G^0_{\mathrm{ion}}$ | standard Gibbs energy of ion formation in aqueous solution, J |
| $\Delta G^0_{\mathrm{a}}$ | standard Gibbs energy of anion formation, J |
| $\Delta G^0_{\mathrm{c}}$ | standard Gibbs energy of cation formation, J |
| $(\Delta G_{\mathrm{s}})_z$ | change of Gibbs energy in formation of an infinitely dilute solution of substance $z$ in water, J |
| $(\Delta G^0_z)_{\mathrm{aq}}$ | standard Gibbs energy of forming substance $z$ in aqueous solution, J |
| $g$ | tons of consumption of raw material, per ton of process end product |
| $H$ | enthalpy (total or specific), J; J/kg, J/mol |
| $\Delta H^0_z$ | standard heat of formation of substance $z$, J |
| $\Delta H^0_{\mathrm{ion}}$ | standard enthalpy of ion formation in aqueous solution, J |
| $\Delta H^0_{\mathrm{a}}$ | standard enthalpy of anion formation, J |
| $\Delta H^0_{\mathrm{c}}$ | standard enthalpy of cation formation, J |
| $(\Delta H_{\mathrm{s}})_z$ | enthalpy change in formation of infinitely dilute solution of substance $z$ in water, J |
| $(\Delta H^0_z)_{\mathrm{aq}}$ | standard heat of formation of substance $z$ in aqueous solution, J |
| $\Delta H^0_{\mathrm{com}}$ | heat of combustion, kJ/kg |
| $I$, $i$ | energy (total and specific), J, kW h; J/kg, J/mol, kW h/ton |
| $K$, $k$ | monetary capital investment |
| $L$ | heat output, kW |

| | |
|---|---|
| $M$ | input and output material flows, ton |
| $m$ | mass of elements, anions, cations |
| $m$ | number of considered energy carriers |
| $N$ | electric capacity, kW |
| $N$ | number of process links in a technological scheme |
| $n$ | amount of process product |
| $n$ | number of conisdered technologies in one link of technological chain |
| $n_e$ | number of electrons in a reaction |
| O | oxygen content in fuel |
| $P$ | energy conserved potential, J, kW h |
| $p$ | pressure, Pa |
| $Q$ | heat, J, kW h |
| $\Delta Q$ | heat losses, J/kg, J/mol, kW h/t |
| $Q_h$ | high calorific value, kJ/kg |
| $Q_l$ | low calorific value, kJ/kg |
| $q$ | heat amount per unit of substance, J/kg, J/mol, kW h/ton |
| $q$ | specific heat consumption in electricity production, g c.e./kW h |
| $R$ | energy conserved, J, kW h |
| $r$ | number of processes in a production scheme |
| $S$, $s$ | entropy (total and specific), J/K; J/kg · K, J/mol · K |
| $\Delta S$ | entropy change, J/K |
| $T$ | temperature in Kelvin, K |
| $T_0$ | temperature of environment, K |
| $t$ | temperature on a Celsius scale, °C |
| $t$ | time, h, y |
| $U$ | internal energy, J |
| $V$, $v$ | volume (total and specific), $m^3$; $m^3$/kg, $m^3$/mol |
| $V$, $v$ | industrial pollutants (total and specific), $m^3$; $m^3$/ton of process product, or ton pollutants per ton of process product |
| $W$ | fuel moisture |
| $W_w$ | electricity production, kW h |
| $w$ | number of seasons in year |
| $Z$, $z$ | costs (total and specific), currency/year; currency/year per unit of process product |
| $\alpha$ | conversion factor of different energy carriers into common units of energy, g c.e./kW h, kg c.e./GJ |
| $\beta$ | fraction of technology in product output specific fuel consumption, g c.e./kW h, kg c.e./GJ |
| $\gamma$ | coefficient of useful product extraction in a process |
| $\delta$ | content of useful component in raw and auxiliary materials |
| $\varepsilon$ | coefficient of cost elements |
| $\varepsilon_s$ | interest of capital |
| $\nu$ | fraction of chemical compound (element) per unit of considered substanc |
| $\varphi_{ion}^{\phi}$ | standard electrode potential, V |

## Subscripts

| | |
|---|---|
| a | anion |
| aq | aqueous |
| ave | average |
| c | cation |
| cc | production of cogeneration plant by condensing mode |
| ch | chemical |
| c | cooling system |
| ck | coke |
| cl | coal |
| compl | complex |
| con | concentrate, concentration |
| cons | consumed |
| cp | cogeneration plant |
| ct | production of cogeneration plant by thermal mode |
| e | electron |
| ec | existing condensation power plants |
| el | electric energy |
| en | energy |
| endo | endothermic |
| enw | electric network |
| es | electrical smelting |
| ex | exergy |
| exo | exothermic |
| f | fuel |
| f | furnace facility |
| h | high |
| hnw | heat network |
| $i$ | additional reference species |
| $i$ | original substance of the process |
| ib | industrial boiler |
| id | ideal |
| if | intersystem power flow |
| ion | ion |
| ir | iron |
| $j$ | element |
| $j$ | process |
| $k$ | final substance of the process |
| $k$ | resulting reference species |
| l | low |
| $l$ | product from the process |
| los | loss |
| $m$ | mechanical |

| | |
|---|---|
| mat | material |
| ml | melting |
| n | nominal |
| nc | new condensation power plants |
| oc | oxygen converter |
| oh | open-hearth |
| or | ore |
| p | heat capacity at constant pressure |
| *p* | product |
| *p* | technology |
| pb | peak boilers |
| pr | previous |
| q | thermal |
| r | raw material |
| res | resulting |
| r | heat recovery installation |
| s | infinitely dilute solution |
| sm | smelting, smelted |
| st | structural change |
| stb | stand-by |
| str | steel as rolled stock |
| t | technological |
| tot | total |
| us | useful |
| w | waste |
| *z* | substance |

## Superscripts

| | |
|---|---|
| abs | absolute |
| add | additional |
| b | base |
| ce | coal extraction |
| ck | coke |
| con | concentration |
| cs | cast |
| d | discharge (industrial discharge) |
| ext | extraction |
| fut | future |
| g | gases (waste gases) |
| id | ideal |
| met | metallurgical |
| opt | optimal |

| | |
|---|---|
| or | ore |
| pel | pellets |
| pr | production |
| qn | quenching |
| real | real |
| rel | relative |
| rol | rolling |
| rs | rolled stock |
| sc | scrap |
| st | steel |
| str | steel as rolled stock |

## Abbreviations

| | |
|---|---|
| CCM | continuous casting machine |
| ESS | energy supply system |
| SER | secondary energy resources |

# Introduction

Chapter 1 briefly describes the concepts and laws of thermodynamics that form the basis for the exergy method. It also considers in detail the technique for drawing and analyzing the complete energy balance of a process, methods for computing the chemical energy and exergy of substances which have been developed with the participation of the author and differ from those proposed by other researchers (Chaps. 2 and 3).

Chapter 4 presents the method of complex analysis of processes and plants which includes both the thermodynamic and techno-economic analysis on the example of utilizing thermal secondary energy resources. Such an analysis allows the determination of both the amounts of secondary energy resources which are theoretically and technically applicable for the utilization and the degree of their economically optimal use.

Despite the fact that the fundamentals of the exergy method of analysis have largely been elaborated, this method has not yet found broad practical application. Therefore, the author has tried to show the practical value of this method, its simplicity and the results that can be obtained.

Chapters 5 and 6 give examples of drawing up the complete energy balance of individual technological processes and complexes of ferrous and non-ferrous metallurgy, determining their efficiency, analyzing their losses with identification of irrevocable losses and those that are theoretically recoverable.

In Chaps. 7 and 8 we estimate the energy reserves of industries (comprised of technological processes) using the usual methodological principles based on the concept that energy conservation in an industry results from more efficient energy use in its individual technological processes, as demanded by the final products.

Thus, the contents of the book can be divided into three sections. The first includes Chaps. 1–4 and can be titled "The theoretical bases for the analysis of energy use in technological processes". The second section (Chaps. 5, 6) is devoted to the application of the developed methods for the analysis of metallurgical processes and technological schemes. The third part (Chaps. 7, 8) illustrates the application of these methods for solving larger problems, such as estimation of energy reserves in industries, forecasting the structure, i.e., the types of fuels used and amounts of future energy consumption.

# 1. The Technological Process as a Subject of Thermodynamic Analysis

All processes proceeding in nature and technical facilities transform some given form of energy into another. Therefore, material changes that underlie any technological process should be considered as the consequences of the energy transformations.

In assessing the thermodynamics of a process, the capability of different processes to selectively utilize energy inputs is evaluated. Only energy of the required kind is consumed while other energy forms are rejected. Because this selectivity is a function of the physical and chemical parameters of a system, to optimize a technological production process the interrelation between material and energy changes, the laws of energy transformation must be understood.

Thus, in the analysis of technological processes it is required to study material transformations and first of all, the processes causing them and the laws of energy transformations. The goal of thermodynamics to devise methods for such studies [13, 28, 38, 67, 69, 87, 89–92, 199, 202].

## 1.1 Thermodynamic Systems and Processes

*A thermodynamic system* is a set of material bodies and fields interacting with each other and with the environment. The choice of the system is purely arbitrary and depends only on the objectives of the experiment. Thermodynamic phenomena can be considered to be macrophysical, independent of the underlying microphysical processes if the dimensions of the studied bodies are large in comparison to the size of elementary particles and the distances between them.

Thermodynamic systems are characterized by the so-called state parameters that are divided into *intensive* and *extensive* (additive). Intensive parameters (such as pressure and temperature) do not depend on the amount of a substance in the system, whereas extensive or additive ones (e.g. volume, heat capacity) are proportional to the system mass.

Interactions between a system and the environment are taken into account only in so far as they are connected with the energy and substance exchange of interest. If there is no energy and substance exchange, the system is called *isolated.* If they interact only due to energy exchange, the system is called *closed.* If interaction between them is represented by both energy and substance exchanges, the system is *open*. Hence, the production processes discussed in this book which are continuously supplied with raw material, fuel, blast air, etc. and from which products and wastes are removed, are open thermodynamic systems.

Interaction between the system and the environment results in changes in the state parameters, i.e., the system does perform a thermodynamic process. These processes are classified as either ***equilibrium*** or ***non-equilibrium***. In the first, the macroscopic parameters of the system are the same at each point in space during each phase of the process. A non-equilibrium process is one where different parts of the system might have different temperatures, pressures, densities, etc.

The concept of an equilibrium process allows a time-independent description of the equilibrium state to study dynamics of processes of macroscopic bodies.

A process is called *reversible*, if as it proceeds forward and backward again the system returns to the initial state, sequentially passing through the same equilibrium states, and no residual changes appear in the system or the environment. Processes that do not satisfy this condition are called *irreversible*.

Irreversible processes include those that require the system to pass through non-equilibrium states and dissipative processes in which the system energy is dissipated to the environment (via friction, diffusion, radiation and heat transfer).

All real processes occurring with macroscopic bodies are irreversible. However, depending on the conditions under which a process occurs, the degree of its irreversibility can be different. Conditions can always be devised so that any technological process can proceed at near-equilibrium, without substantial energy losses due to friction, thermal conductivity, diffusion and radiation. This is achieved by using lubricants, polished surfaces, heat-insulating materials and other measures. The degree of irreversibility can be made so low that a process can be treated as practically reversible.

Thus, the limiting case for any real irreversible process is the appropriate reversible process, the analysis of which allows one to judge the optimum possible conditions for a real production process. This allows the application of the methods of classical thermodynamics which, strictly speaking, should be used only for equilibrium and reversible processes.

## 1.2 The Laws of Thermodynamics

Thermodynamics is based on three laws that have been formulated on the basis of a generalization of a large number of physical and chemical data. The laws relate changes in system parameters to changes in the system state. The First Law is a formulation of the law of conservation of energy and energy transformation. The Second Law characterizes the direction of macroscopic processes and determines the conditions for mutual transformation of work and heat. The Third Law (thermal theory of Nernst) has an important, though more limited significance and can be used for the determination of a numerical absolute value for one of the most important thermodynamic functions, entropy. The substance of these laws and some of the more important thermodynamic functions are described briefly below.

### 1.2.1 Internal Energy, Work and Heat. The First Law of Thermodynamics.

In general, any thermodynamic system moves with respect to the surrounding bodies and is characterized by some kinetic and potential energy. Furthermore, the internal microstructure of a system, the ensemble of numerous particles (molecules, atoms, etc.), also carries energy when the particles perform various motions. The total energy of all internal motions in a system is called its internal energy $U$. This is generally the quantity of interest rather than the overall system energy. The internal energy is a function of the state of the system. Its change does not depend on the nature of the process and is determined by the initial and final states of the system only. Mathematically, this means that the differential of internal energy is a total differential.

The internal energy can include the kinetic energy of motion of some macroscopic parts of the system with respect to others, the energy of the random motion of atoms and molecules, the energy of chemical and phase transformations, the energy of intramolecular and intraatomic motions, the energy of nuclear forces, the internal energy of elementary particles, etc. This list is nevertheless incomplete. The absolute value of the internal energy of any macroscopic body cannot be determined. This is true of all thermodynamic functions except for the entropy whose absolute value can be found. Thus, only the change in the internal energy of a macrobody can be measured accurately, usually, relative to a previously defined standard state.

Heat and work do not refer to properties of the system, their change depends on the nature of the thermodynamic process. Although they are both characteristics of the interaction between a system and its environment, they are very different parameters. Heat is the microscopic energy exchange between the system and environment due to the random (thermal) motion of the microscopic particles. Work characterizes the energy change at a macroscopic level. It is related to the directed displacement of macroscopic bodies. The differentials of heat and work are partial differentials.

In thermodynamics work is defined as in mechanics: $dA = \sum_i P_i dl_i$, i.e., as the sum of the products of the forces $P_i$ acting on the system and the elementary displacements $dl_i$ of the system in the directions of the forces. Some components of this general expression for work can be caused by the action of diverse forces (mechanical, electric and magnetic forces, forces of chemical affinity, etc.).

In many practical cases work of mechanical forces is considered in continuous media. The total action of mechanical forces is expressed by the pressure $p$ with which the system acts on the environment or the environment on the system, resulting in a change of volume $dV$. This mechanical work of compression or expansion is determined from the relation

$$dA_V = pdV \ . \tag{1.1}$$

If other forces are present, the expression is generalized to

$$dA = pdV + dA^* \qquad (1.2)$$

where $dA^*$ is the non-mechanical work caused by the action of generalized external forces in addition to the environmental pressure.

If a closed thermodynamic system interacts with its environment, heat exchange occurs in the form of heat and work, resulting in a transition of the system to some other equilibrium state. The difference between the heat $Q$ absorbed by the system and the work $A$ done with respect to the environment is the same for any transition from the initial to the final state, i.e., it does not depend on the nature of the process.[1] This difference determines the change in the systems's internal energy $\Delta U$ which cannot depend on the nature of the process (due to the law of energy conservation $U$ = const) and is a function only of the initial and final states.

Thus, for the general case, we can write $\Delta U = Q - A$ or for an infinitely small change in the system energy

$$dU = dQ - dA \qquad (1.3)$$

All thermodynamic – both reversible and irreversible – processes are governed by the first law of thermodynamics. It holds for closed and open systems. Note, however, that in open systems the internal energy changes due to the heat released and work done as well as due to the energy contained in matter exchanged with the environment. In other words, for open systems it is necessary to consider the additional factor of the work done by the flow of matter within the studied system. Hence, in this case the equation for the first law (1.3) should include an additional term that takes into account the overall change of kinetic and potential energy due to mass flow, mechanical work of compression or expansion, work done by the flow or supplied to it and the work against frictional forces.

The first law of thermodynamics incorporates all the properties necessary to construct an energy balance of some production process. However, it does not provide any information on the possibility of its course. Not all processes can in fact be implemented. The answer to this question lies in the second law of thermodynamics.

### 1.2.2 The Second Law of Thermodynamics

Like the first law, the second law of thermodynamics is formulated on the basis of numerous experiments. According to the first law, mutual conversion of thermal and mechanical energy is strictly equivalent $A \rightleftarrows Q$. However, the forward and reverse transitions are not the same process. Conversion of mechanical energy to

[1] In thermodynamics, heat absorbed by the system is a positive quantity and that released is negative. Work done by the system with respect to the environment is positive and that done by external forces with respect to the system is negative.

thermal energy proceeds spontaneously. To execute the reverse process of converting heat into work, it is necessary to have a heater, producing heat $Q_1$ and a cooler, absorbing heat $Q_2$. Hence, we can write

$$A = Q = Q_1 - Q_2$$

The second law of thermodynamics states the impossibility to create an engine that completely (without compensation) transforms heat into work (S. Carnot). Other formulations of the second law are known.

Without dealing with the mathematical formulas we note an important corollary from the second law. The value $dQ/T$ remains constant during any reversible transitions between two fixed states of the system. This means that this quantity is a state function with the total differential

$$dS = dQ/T \tag{1.4}$$

This function is called *entropy*. Hence, at $T = \text{const}$ we have

$$Q = T(S_2 - S_1)\ . \tag{1.5}$$

In irreversible processed $dQ_{\text{ir}} < dQ$. Thus, taking into account (1.4)

$$dS > dQ_{\text{ir}}/T\ . \tag{1.6}$$

Relations (1.4) and (1.6) represent a general analytic formulation of the second law: for infinitesimal reversible changes in the state of a system which are accompanied by absorption of a quantity of heat $dQ$ by the system, the differential of the system entropy is equal to the ratio $dQ/T$. For infinitesimal irreversible changes in the state of the system which are accompanied by the absorption of heat by the system $dQ_{\text{ir}}$, the differential of the system entropy always exceeds the ratio $dQ_{\text{ir}}/T$.

### 1.2.3 The Third Law of Thermodynamics

The second law of thermodynamics allows one to determine only changes in the system entropy that correspond to reversible transitions from some thermodynamic equilibrium state to another. This is quite sufficient for understanding the thermodynamics of processes which do not result in the production of new substances. Then the entropies of the initial state $S_1$ and the final state $S_2$ refer to the same species comprising the system, and the additive constant resulting from integration of $dS = dQ/T$ can be neglected.

Chemical thermodynamics is concerned with the formation of new chemical substances that were initially absent in the system. Here $S_1$ refers to the initial substances and $S_2$ to the final ones. In this case it is necessary to know the absolute values of the entropy.

Experimental studies by W. Nernst and other scientists showed that during unlimited equilibrium cooling of macroscopic bodies the entropy diminishes and tends to the minimum possible value which does not depend on any other factors such as pressure, volume, magnetic or other field. Nernst formulated his "thermal theorem" as follows: "Near absolute zero, the entropies of all the substances in equilibrium with one another are constant and equal to one another."

Later Planck showed that at $T = 0$ the entropies are indeed equal to one another; they are zero. Planck presented a more general formulation of Nernst's law: "The entropies of all bodies contained within a finite volume of homogeneous composition decrease toward a temperature tending to absolute zero."

Thus, the third law of thermodynamics can be written in the form

$$\lim_{T \to 0} S(T, V, \ldots) = 0 \tag{1.7}$$

The third law is of great practical importance since it allows the absolute values of entropy to be calculated. Directly from the general formulation of Nernst's theorem it follows that the additive constants of entropy are equal to one another and are also zero. Thus, for any process we can calculate the absolute entropy. For an isobaric (constant pressure) process, for example, it is determined from

$$S = \int_{T=0}^{T} (C_p/T)\, dT . \tag{1.8}$$

where $C_p = f(T)$ is the dependence of the heat capacity on the temperature.

## 1.3 State Functions

The Gibbs method of thermodynamic functions of state is the most general and inclusive method for studies on diverse systems and processes. It can be applied to any system at thermodynamic equilibrium and to any process that can be considered as equilibrium and reversible. By certain further assumptions the Gibbs method can also be applied to non-equilibrium processes. This will not be discussed here.

The main thermodynamic state functions are internal energy $U$, entropy $S$, enthalpy $H$, free energy $F$, free enthalpy $G$. The first two functions have been considered earlier. Enthalpy represents the sum of internal energy $U$ of the system and the product of the system pressure $p$ and volume $V$, i.e.,

$$H = U + pV . \tag{1.9}$$

Since the new function is a combination of state functions $U$, $p$, $V$, it is also a state function.

From the first law of thermodynamics, for a system doing work of expansion only, $dQ = dU + pdV$, and using the Legendre transformation $pdV = d(pV) - Vdp$ we obtain

$$dQ = d(U + pV) - Vdp = dH - Vdp . \tag{1.10}$$

Thus it follows that in an isobaric ($dp = 0$) process the heat input to the system can change only the enthalpy, i.e., $dQ = dH$. Therefore, enthalpy can be used to analyze processes occurring in near-isobaric conditions. The state function $F = U - TS$ is the *free energy* (the *Helmholtz energy*) and $G = H - TS$ is the free enthalpy (the *Gibbs energy*). These functions determine the useful work done by a system.

From the first and second laws of thermodynamics the elementary useful (non-mechanical) work that can be done by a system in a reversible process is

$$dA^* = -dU + TdS - pdV \tag{1.11}$$

For an isochoric-isothermal process ($V = \text{const}$, $T = \text{const}$) this relation is

$$dA^*_{V,T} = -d(U - TS) = -dF \tag{1.12}$$

or, for finite changes of the system state it has the form

$$A^*_{V,T} = -\Delta F\ . \tag{1.13}$$

Because the work delivered by a system in an irreversible process is always less than the work of a reversible process $A_{ir} < A$, one can conclude that the maximum non-mechanical work of a system in an isochoric-isothermal process is equal to the decrease of the system's free energy.

The behavior of isobaric-isothermal systems is conveniently described by the free enthalpy. In this case ($p = \text{const}$, $T = \text{const}$) expression (1.11) will have the form

$$dA^*_{p,T} = -d(U - TS + pV) = -dG \tag{1.14}$$

or

$$A^*_{p,T} = -\Delta G \tag{1.15}$$

for finite changes in the system. Relations (1.14, 15) mean that the maximum useful (non-mechanical) work of a system for isobaric-isothermal processes is equal to the decrease of the system's free enthalpy.

These thermodynamic functions are, of course, interrelated. For example, the system's internal energy can be expressed via the enthalpy as follows

$$U = H - pV \tag{1.16}$$

$$U = F + TS \tag{1.17}$$

$$U = G + TS - pV\ . \tag{1.18}$$

Sequential substitution of the obtained expressions for internal energy in the thermodynamic equation (1.11)[2] gives three additional equations:

[2] The thermodynamic equation for systems doing only mechanical work, i.e., equation (1.11) at $dA^* = 0$.

$$dH = TdS + Vdp \tag{1.19}$$

$$dF = -SdT - pdV \tag{1.20}$$

$$dF = -SdT + Vdp \tag{1.21}$$

which are the total differentials of enthalpy, free energy and free enthalpy for systems doing only work of expansion.

The following equalities

$$T = \left(\frac{\partial U}{\partial S}\right)_V ; \quad p = -\left(\frac{\partial U}{\partial V}\right)_S ; \quad T = \left(\frac{\partial H}{\partial S}\right)_p ; \quad V = \left(\frac{\partial H}{\partial p}\right)_S \tag{1.22}$$

$$p = -\left(\frac{\partial F}{\partial V}\right)_T ; \quad S = -\left(\frac{\partial F}{\partial T}\right)_V ; \quad V = \left(\frac{\partial G}{\partial p}\right)_T ; \quad S = -\left(\frac{\partial G}{\partial T}\right)_p$$

follow from the differentials of $U$, $H$, $F$, $G$ (at $A^* = 0$)

$$dU = TdS - pdV ; \qquad dH = TdS + Vdp ;$$

$$dF = -SdT - pdV ; \qquad dG = -SdT + Vdp .$$

For such systems the functional dependences $U(S,V)$; $H(S,p)$; $F(T,V)$; $G(T,p)$ are called *characteristic functions* of the system. Since the difference in these functions between two states is the maximum useful work done by the system in a reversible transition, in analogy to mechanics, these functions are also called *thermodynamic potentials*. Substituting $p$, $V$ and $S$ in the expressions for $U$, $H$, $F$ and $G$ with respect to equations (1.22) we obtain

$$U = H - pV = H + V\left(\frac{\partial U}{\partial V}\right)_S ; \tag{1.23}$$

$$H = U + pV = U + p\left(\frac{\partial H}{\partial p}\right)_S ; \tag{1.24}$$

$$F = U - TS = U + T\left(\frac{\partial F}{\partial T}\right)_V ; \tag{1.25}$$

$$G = H - TS = H + T\left(\frac{\partial G}{\partial T}\right)_p \tag{1.26}$$

Equations (1.25, 26) are called the *Gibbs-Helmholtz equations*.

The thermodynamic potential $\varphi_{\text{ch}}(p,T)$ is the *Gibbs energy* of a homogeneous system per unit mass and is called the *chemical potential*

$$\varphi_{\text{ch}} = i - sT \tag{1.27}$$

where $i$ and $s$ are the specific (per unit) enthalpy and entropy of the system.

The chemical potential $\varphi_{\text{ch}}$ is the partial derivative of any thermodynamic potential of the system of mass $M$ at constant values of the independent variables, i.e.,

$$\left(\frac{\partial U}{\partial M}\right)_{V,S} = \left(\frac{\partial H}{\partial M}\right)_{p,S} = \left(\frac{\partial F}{\partial M}\right)_{V,T} = \left(\frac{\partial G}{\partial M}\right)_{p,T} = \varphi_{\text{ch}} \tag{1.28}$$

For chemical reactions, the chemical potential is the maximum useful work that can be done by a reacting body with respect to an external object when the body mass changes by a single unit.

## 1.4 Thermodynamic Properties of Substances and Their Changes in Chemical Processes

Physico-chemical processes are taken by us as phase transformations (melting, evaporation, crystallization etc.), formation of solutions; processes occurring on surfaces interfacing particular phases, such as, for example, adsorption; processes of appearance and growth of a new phase (for example, appearance and growth of crystals). Chemical processes include the formation of new chemical substances that were initially absent in the systems.

As was noted earlier, the values of energy input and output and the values of useful work done by the system are the most important results of changes in a system state from a practical viewpoint. They can be determined by calculating the value of changes in the thermodynamic potential, i.e, the difference in the values of this potential in the initial and final system states. As was discussed above, the third law of thermodynamics states that only entropy can be determined absolutely. Functions of state, that is, internal energy, enthalpy, free energy and free enthalpy, however, can be found with respect to some reference state $U_0$, $H_0$, etc. Then

$$\Delta U = U_2 - U_1 = (U_2 - U_0) - (U_1 - U_0) \ , \tag{1.29}$$

$$\Delta H = H_2 - H_1 = (H_2 - H_0) - (H_1 - H_0) \ . \tag{1.30}$$

The choice of the reference state is not dictated by any constraints and is only a factor of practical convenience. However, once a particular reference state is chosen, it must be applied to all elements of the system.

The definition of reference states is fixed by international agreement. In the absence of external force fields the state of a system is defined by the pressure and temperature or the volume and temperature. Since many systems are open to the atmosphere, constant pressure states are perhaps of more practical importance. Therefore, the more frequently used state functions are heat capacity $C_p(p,T)$, entropy $S(p,T)$, enthalpy $H(p,T)$, free enthalpy (the Gibbs free energy) $G(p,T)$. The reference state is the thermodynamic equilibrium state at a pressure of one atmosphere (760 mm mercury or 101325 Pa) and a temperature of either 0 K or 298.15 K (−273.15°C or +25°C).

Thermodynamic properties at a constant pressure of one atmosphere are functions of temperature and are called *standard thermodynamic functions*. The values of thermodynamic properties at any other pressure $p$ are expressed through the standard thermodynamic functions with corrections for the change in pressure. These corrections are given

$$\delta C_p = C_p(p,T) - C_p^0(T) = -\int_{p_0}^{p} T\left(\frac{\partial^2 V}{\partial T^2}\right)_p dp\,; \tag{1.31}$$

$$\delta S = S(p,T) - S^0(T) = -\int_{p_0}^{p} \left(\frac{\partial V}{\partial T}\right)_p dp\,; \tag{1.32}$$

$$\delta H = H(p,T) - H^0(T) - \int_{p_0}^{p} \left(V - T\left(\frac{\partial V}{\partial T}\right)_p\right) dp\,; \tag{1.33}$$

$$\delta G = G(p,T) - G^0(T) = \int_{p_0}^{p} \left(V - T\left(\frac{\partial V}{\partial T}\right)_p\right) dp + T\int_{p_0}^{p} \left(\frac{\partial V}{\partial T}\right)_p dp\,. \tag{1.34}$$

In chemical reactions the external energy, entropy, heat capacity or free enthalpy of substances formed differ, as a rule to a greater or lesser extent from the values of the appropriate quantity of initial substances that are considered at the same temperature and pressure, as final substances. The difference between the corresponding values of final products and initial substances at the constant temperature and pressure is called enthalpy, entropy, heat capacity or free enthalpy of the appropriate reaction at the given values of temperature and pressure

$$\Delta H_r(p,T)\,; \quad \Delta S_r(p,T)\,; \quad \Delta C_{p_r}(p,T)\,; \quad \Delta G_r(p,T)\,.$$

From the definition of free enthalpy we have

$$\Delta G_r(p,T) = \Delta H_r(p,T) - T\Delta S_r(p,T)\,, \tag{1.35}$$

which is the fundamental equation of chemical thermodynamics.

## 1.5 Thermochemistry

Chemical reactions are accompanied by a change in the internal energy of the reacting system that manifests itself as the release or absorption of heat by this system. Reactions proceeding with heat release are called *exothermic*, and those with heat absorption are *endothermic*. The heat released or absorbed during a chemical reaction proceeding at constant volume or constant pressure and measured when the temperature of the products and reactants is the same is called the heat effect of a reaction $\Delta H$.[3] Thus, each reaction can be characterized by the isochoric $Q_{V,T}$ or isobaric $Q_{p,T}$ heat effect. According to the first law of thermodynamics this heat effect is equal to the change in the internal energy or enthalpy of the reacting system at the temperature $T$, respectively,

[3] The heat of an exothermic reaction is taken as positive and that of an endothermic reaction negative.

$$Q_{V,T} = \Delta U(V,T) \tag{1.36}$$

$$Q_{p,T} = \Delta H(p,T)\ . \tag{1.37}$$

The relation between the enthalpy and internal energy of an arbitrary system yields a similar relation between their changes at the temperature $T$:

$$\Delta H(p,T) = \Delta U(V,T) + \Delta(pV)_T\ , \tag{1.38}$$

where $\Delta(pV)_T$ is the difference in the product $pV$ of the reacting system at the beginning and at the end of the reaction at the same temperature $T$.

As was noted above, owing to the greater practical importance of chemical reactions proceeding at a constant pressure, only isobaric processes will be dealt with below.

The isobaric heat effect $Q_{p,T}$ of an arbitrary reaction can be expressed through the molar enthalpy $H_{\text{mol}}$ of the reactants and final products in the following way:

$$Q_{p,T} = \Delta H_{\text{r}}(T) = \sum_{i=1}^{m} y_i H_{\text{mol}}(Y_i,T) - \sum_{k=1}^{n} x_k H_{\text{mol}}(X_k,T) \tag{1.39}$$

where $\Delta H_{\text{r}}(T)$ is the reaction enthalpy, $X_1, X_2, \ldots, X_n$ are the reactants, $Y_1, Y_2, \ldots, Y_m$ the final products, $x_1, x_2, \ldots, x_n$ the stoichiometric coefficients of the reactants and $y_1, y_2, \ldots, y_m$ the stoichiometric coefficients for the final products.

Relation (1.39) is an analytical expression of the *Hess law* that says "the heat effect of a chemical reaction depends on the initial and final states of substances in the reaction and does not depend on the intermediate stages and mechanism of the reaction itself." This law was discovered by the Russian chemist G. I. Hess before the first law of thermodynamics and is the principal law of thermochemistry.

One of the corollaries of the Hess law is the law formulated by the German physicist Kirchhoff on the dependence of the heat effects of a reaction on the temperature. It says that the heat effects of enthalpy $\Delta H_{\text{r}}(T)$ and $\Delta H_{\text{r}}(T_0)$ of the same reaction at the temperatures $T$ and $T_0$ and constant pressure are expressed by

$$\Delta H_{\text{r}}(T) = \Delta H_{\text{r}}(T_0) + \Delta\,[H(T) - H(T_0)]\ , \tag{1.40}$$

where

$$\Delta[H(T)-H(T_0)] = \sum_{i=1}^{m} y_i[H(T)-H(T_0)]_{Y_i} - \sum_{k=1}^{n} x_k[H(T)-H(T_0)]_{X_k} \tag{1.41}$$

Relations (1.40, 41) show that in order to calculate the heat effect of a reaction at any temperature it is necessary to know its value at some temperature $T_0$ and the changes of the enthalpy $H(T) - H(T_0)$ of all the components of the reaction.

Another important corollary of the Hess law is that the heat effect (enthalpy) of any reaction is equal to the difference between the sums of the enthalpies of formation of the final products and the sum of the enthalpies initial reactants

$$\Delta H_r(T) = \sum_{i=1}^{m} y_i \Delta H(Y_i, T) - \sum_{k=1}^{n} x_k \Delta H(X_k, T) \tag{1.42}$$

where $\Delta H(X_k, T)$ is the enthalpy of formation of the compound $X_k$ from its atomic elements.

Thus, the heat of an arbitrary chemical reaction can be expressed by (1.42) through the enthalpy of forming the reaction components. Therefore, the fundamental thermochemical characteristic of a chemical substance in a given phase state is its enthalpy of formation from atomic elements.

The *standard state* of chemical elements is their state under standard conditions, are called their *standard phase states*. Under these standard phase states of chemical elements: oxygen, hydrogen, nitrogen, flourine, and chlorine ($O_2$, $H_2$, $N_2$, $F_2$, $Cl_2$) are diatomic gases, graphite $(C)_s$ is a crystalline solid, bromine $(Br_2)_l$ is liquid, iodine $(I)_s$ is solid, sulphur is solid (crystalline orthorhombic) $S_{c\ orth}$; rare gases He, Ne, Ar, Kr, Xe, Rn are gas. All other elements except for mercury are crystalline. The standard enthalpies of formation ($p_0 = 101325$ Pa, $T_0 = 298.15$ K) are tabulated in numerous sources [32, 43, 65, 85, 100, 101, 126, 128, 133, 171, 189].

## 1.6 Maximum and Minimum Work. The Gouy-Stodola Law

As was shown above, the work done by a system with respect to external objects (the environment) in a reversible transition from equilibrium state 1 to equilibrium state 2 is maximal. It is important to note that the work done by the environment to return the system from state 2 to state 1 by the same path will be minimal. It is evident that

$$A_{max} = -A_{min} \,. \tag{1.43}$$

Both these ideas are very important in the analysis of real technical systems. In fact, the maximum work can characterize the utmost level of energy use and the minimum work can determine the minimal value of the energy (work) required.

Of practical concern is, of course, the maximum useful work that can result from the reversible transition of a system (source of work) from an arbitrary equilibrium state 1 to a state of thermodynamic equilibium with the environment that has an invariable temperature $T_0$ and pressure $p_0$.

An aggregate system, consisting of the environment and a source of work, can be treated as an isolated non-equilibrium system and, hence, can do work. We shall try to determine the maximum (useful) work of this system.

We denote the internal energy, pressure, temperature and volume of a work source by $U_1$, $p_1$, $T_1$ and $V_1$; $p_2 = p_0$, $T_2 = T_0$ and $V_2$ in the initial and final states, respectively. $U_{0_1}$ and $U_{0_2}$ stand for the internal energy of the environment in the initial and final states, respectively.

The total internal energy of the combined system in the initial non-equilibrium state will be equal to $U'_{syst} = U_1 + U_{0_1}$, and in the final equilibrium state they are $U''_{syst} = U_2 + U_{0_2}$. In an isolated system work can be done only at the expense of a decrease in its internal energy, i.e.,

$$A = U'_{syst} - U''_{syst} = (U_1 - U_2) + \left(U_{0_1} - U_{0_2}\right) \tag{1.44}$$

But in a combined non-equilibrium system interaction between the work source and the environment can, in general, result in a work $A_0$ and heat transfer $Q_0$.

In accordance with the first law of thermodynamics

$$U_{0_2} - U_{0_1} = Q_0 + A_0 \ . \tag{1.45}$$

The work source can perform work with respect to the environment (against its pressure) which equals[4]

$$A_0 = p_0(V_2 - V_1) \tag{1.46}$$

Substituting the expression obtained for work in (1.45) we have

$$U_{0_1} - U_{0_2} = -Q_0 - p_0(V_2 - V_1) \ . \tag{1.47}$$

Heat transferred from the work source to a constant-temperature environment will be equal to

$$Q_0 = T_0\left(S_{0_2} - S_{0_1}\right) \tag{1.48}$$

Substituting (1.47, 48) into (1.44) we have

$$A = (U_1 - U_2) - T_0\left(S_{0_2} - S_{0_1}\right) - p_0(V_2 - V_1) \ . \tag{1.49}$$

The resulting equation gives the value of useful work that can be done by an isolated system during its transition from a non-equilibrium to an equilibrium state, since a part $p_0(V_2 - V_1)$, which is expended for medium compression and hence cannot be used at our will, is subtracted from the total possible work. Equation (1.49), however, does not determine the value of maximum useful work, since its derivation was not based on the conditions of reversibility of processes occurring in the system.

As long as the entropy of an isolated system does not change during reversible processes, we can wirte

$$S_{0_2} - S_{0_1} = S_1 - S_2 \ .$$

From this relation and (1.49) we can derive an expression for the maximum useful work done by an isolated system during a transition from a non-equilibrium to an equilibrium state:

---

[4] In process $1 \rightarrow 2$ the environment does work with respect to the work source equal to $p_0(V_{0_2} - V_{0_1})$. Since for the isolated system $V + V_0 = \text{const}$, then $p_0(V_{0_2} - V_{0_1}) = -p_0(V_2 - V_1)$.

$$A_{max} = (U_1 - U_2) - T_0(S_1 - S_2) - p_0(V_2 - V_1) . \quad (1.50)$$

In order to determine the maximum work, (1.45) must take into account the work done by the environment during a change in the state of the system equal to $p_1 V_1 - p_0 V_1$. Then we obtain

$$A_{max} = (U_1 - U_2) - T_0(S_1 - S_2) - p_0(V_2 - V_1) + p_1 V_1 - p_0 V_1 , \quad (1.51)$$

or denoting the system volume, enthalpy and entropy in the equilibrium state with the environment by $V_0$, $H_0$, $S_0$ we get

$$A_{max} = (H_1 - H_0) - T_0(S_1 - S_0) . \quad (1.52)$$

Z. Rant proposed to call this the *exergy* (Sect. 1.7). The value of the useful work that can be done by a mechanical system in real, generally irreversible processes is less than the maximum work by some positive value. In fact, if (1.49) determines the useful work done by an isolated system during any processes, including irreversible ones, and (1.50) determines the same work for reversible processes, then taking into account that $S_{0_2} - S_{0_1} > S_1 - S_2$, we can write

$$\Delta A = A_{max} - A = T_0 \left[ (S_{0_2} - S_{0_1}) - (S_1 - S_2) \right] = T_0 \Delta S^* , \quad (1.53)$$

where $\Delta S^* = [(S_{0_2} - S_{0_1}) - (S_1 - S_2)]$ is the increase in entropy for a combined system due to irreversible processes, $\Delta A$ are the losses of work.

Hence, the larger the increase in the system entropy, i.e., the higher the degree of irreversibility of the processes, the less work done by the system. Equation (1.53) is called the *Gouy-Stodola equation.*

Equations (1.52) and (1.53) are the fundamental and most general relations between the actual and theoretical maximal useful work and hold for any thermodynamic systems.

## 1.7 The Concept of Exergy. The Exergy Method of Analysis

Diverse forms of energy can be transformed into work and other forms of energy in different ways. Electric and mechanical energies can be completely converted into work, whereas heat and the internal energy of a body can only partly be converted into work. The limited convertibility of different energy forms originates from the second law of thermodynamics, by which all processes are irreversible. Hence, the energy forms used in practice can be divided by the extent of their convertibility into three kinds:

1. unlimitedly convertible (kinetic, potential, electric, magnetic, etc.);
2. limitedly convertible (internal energy of bodies and systems that are not in thermodynamic equilibrium with the environment; chemical energy of substances; heat at $T \neq T_\Phi$);

3. unconvertible (internal energy of components of the environment and systems that are in thermodynamic equilibrium with it; heat at $T = T_0$; mechanical energy equal to the work needed to overcome atmospheric pressure).

In practice the unlimitedly convertible energy forms are obviously preferable and more valuable than those whose convertibility is restricted by the second law of thermodynamics. Hence, it is quite natural that work, which is a general indicator of convertibility, was applied by researchers as a performance index of different energy forms.

The notion of the availability a working body appeared in the 19th century. However, it was known only to a few experts in thermodynamics, had different names and was not used in practice. This notion as well as the tools of classical thermodynamics were primarily applied to closed thermodynamic systems [74, 170]. However, technological progress resulted in considerable complication, of the studied objects, in the emergence of new processes referred to a class of open thermodynamic systems. The classical thermodynamic tool (the Clausius method of cycles) turned out to be insufficient and in some cases absolutely unsuitable to analyze such processes. This is because the method of cycles was not intended for evaluating the availability of a material nor for calculating the energy flows entering and leaving a system when the state of the environment changes.

In the first half of the 20th century the method of the Gibbs potentials was widely applied in the analysis of open thermodynamic systems. This method is more rigorous from the logical and mathematical points of view and at the same time more simple. The method is based on classical thermochemistry. It is suitable primarily for the analysis of production processes, by studying chemical reactions describing a process. By definition, thermodynamic potentials are functions of parameters of the studied system only. For these reasons they do not allow one to assess the availability of a system for different states of the environment.

This problem was addressed in the early 1950s, when Z. Rant introduced the new concept of *exergy*. It characterizes the maximum work that can be done by a system in a reversible interaction with its environment, as a result of which all its parameters will equilibrate with the environmental parameters. The term "exergy" is now generally accepted [1–3, 7–14, 25–31, 33–37, 51–53, 55–57, 61–64, 67–72, 95, 109–117, 120–122, 129–132, 147]. Exergy takes into account the first and second laws of thermodynamics: work can be done only if the system is not in equilibrium with its environment; maximum work can be achieved only in reversible processes.

Assuming that energy forms can be conventionally represented as a sum of two qualitatively different parts, unlimitedly convertible energy and energy that cannot be converted into work, Rant called the first component *exergy* and the second *anergy*. Hence, any energy can be broken down into exergy and anergy, in some cases one of them can be equal to zero.[5] Exergy can be converted into

[5] Some authors [34] consider that the determination of energy as the sum of exergy and anergy is in general incorrect and holds only for a heat flow when the temperature $T$ is higher than the environmental temperature, i.e., $T > T_0$.

any form of energy and, therefore, also into anergy, but a reverse transition of anergy into exergy is impossible.[6] Any irreversibility results from the conversion of exergy into anergy.

The accepted expression of energy loss should not be taken literally since it contradicts the law of energy conservation. What is meant is that in the course of its conversion into other forms of energy, some energy is transformed into less valuable forms. Thus, the energy loss is actually the loss of exergy. The law of exergy devaluation says that the sum of input exergies decreases during the process. In irreversible processes exergy is lost irretrievably, whereas energy does not disappear. It simply changes to other, less valuable forms. In other words, in all processes, it is exergy, rather than energy, which is consumed. Exergy is converted into anergy.

The exergy method of analysis is a convenient tool for studying different technical systems and production processes. For example, it allows a more correct account of the quality of the consumed and produced energy in calculations of the energy balance. Thus, a more correct evaluation of the energy consumption can be made for qualitative analysis of the energy losses and for identification of measures to be taken to limit them.

The exergy method, used to study industrial production processes, can determine the current levels of efficiency and assess the maximum possible efficiency. It can show how to decrease the volume of energy resources consumed or to change their composition.

At present the exergy method is widely used. Among the more important publications discussing the subject are [31, 80, 81, 112–114, 174–176, 179, 184–186]. Jeschar and co-authors [31, 80, 81] consider the problem of drawing up an analyzing the energy balance for a ferrous metallurgy plant on the basis of exergy as heat, heat flows and chemical reactions. The concept of exergy makes the energy balance calculated by the general method more exact and clear, as is seen from the comparison made by Jeschar. However, the application of exergy to chemical reactions rather than to chemical substances does make the calculations more complicated and laborious.

In analogy to [31, 80, 81] other authors [111, 136, 137, 195] also use the exergy of chemical reactions to analyze production (chemical) processes.

The works of J. Szargut, who proposed a method for calculating the energy and exergy of substances, proved to be a step forward in developing the method for drawing up energy balances. The energy and exergy of chemical compounds calculated by him can be used just as *calorific value* and *heat of combustion* in chemistry and energetics [173–187]. Thus there is no need to write out all the chemical reactions proceeding in a given process to determine the amount of energy (heat) released and used. This considerably simplifies and streamlines the calculations.

[6] A process in which entropy decreases is an exception. Such a process has an exergy higher than its energy, i.e., anergy attracted from the environment is converted into exergy.

# 2. Efficiency of Technological Processes Based on Energy Balance

The law of conservation of energy, the first law of thermodynamics, allows us to evaluate the efficiency of a given technological process via an energy balance. The methods of drawing the energy balance and the possibilities it affords for studying various technical systems can differ. This author has developed his own version for studying processes with complex physico-chemical transformations of matter and energy.

## 2.1 Heat Balance of a Process

Energy balance based on the first law of thermodynamics is correctly called heat balance, because all energy forms are converted into heat according to the corresponding physical equivalents [5, 15, 16, 21, 44 45, 58, 66, 87, 97–99, 103, 139, 142, 149]. Such a balance is usually written

$$\Sigma Q_{\mathrm{f}} + \Sigma Q_{\mathrm{el}} + \Sigma Q_{\mathrm{r}} + \Sigma Q_{\mathrm{exo}} = \Sigma Q_{\mathrm{p}} + \Sigma Q_{\mathrm{w}} + \Sigma Q_{\mathrm{endo}} + \Sigma \Delta Q \tag{2.1}$$

where $Q_{\mathrm{f}}$ is the heat of fuel combustion; $Q_{\mathrm{el}}$ the heat equivalent to the electric (mechanical) energy supplied for the process; $Q_{\mathrm{r}}$ – sensible heat of the raw material, fuel, blast, etc.; $Q_{\mathrm{p}}$, $Q_{\mathrm{w}}$ – sensible heat of products and wastes; $Q_{\mathrm{exo}}$, $Q_{\mathrm{endo}}$ – heat of exo- and endothermic reactions, respectively; $\Delta Q$ – heat losses from the external surfaces of the aggregate due to e.g., radiation.

As is seen from (2.1), this balance is based on the principle of the heat of the primary chemical reactions occurring in the studied process. Since as a rule, it is difficult to uniquely determine the set of independent chemical reactions for the plausible description of complex production processes, calculations of energy balances using (2.1) do not give comparable results even for similar conditions [153].

In principle, when using (2.1), it is not necessary to include all the reactions which occur. Using the Hess law their total heat can be calculated from the chemical composition of the input and output substances. The resulting heat of all the reactions in a process $Q_{\mathrm{res}}$ that should be included into (2.1) is

$$Q_{\mathrm{res}} = \sum_k \Delta H_k^0 - \sum_i \Delta H_i^0 \tag{2.2}$$

where $\Delta H_0$ is the standard enthalpy of the substance formation, and $i, k$ are the indices of the input and output substances.

But this method of determining the total amount of heat in a process due to chemical reactions does not eliminate the noted drawbacks of the classical energy (heat) balance in practical applications. Calculations of the heat balance in actual processes using this method reveal that it is usually difficult to strike a balance between the left and right sides of (2.1). A large discrepancy amounting to 15–20 % of the total energy input appears. This can be partially explained by the following. It is difficult and sometimes impossible to determine the enthalpies of all the substances produced in metallurgical and chemical processes, as the reference books have no values of $\Delta H$ for slags, mattes and similar materials whose composition may be very complex or unknown. In addition the enthalpies of formation of these substances are usually considerably higher than the values of $\Delta H$ of the initial fuel, raw material, fluxes, etc. Since most processes take place in open systems, even when the amounts of raw materials are carefully measured before processing, afterwards it is difficult to ascertain how much of the product has, for example, been incorporated into the reactor vessel wall. Thus, the accuracy with which the amounts of products are determined is considerably lower than for the reactants. It is lower still for the undesired products such as flue gases, dusts, slags, mattes. Thus, the first term in (2.2) can be determined with less certainty than the second one; this influences the accuracy of the value of the input part of the balance.

But this is only one drawback of the classical (heat) balance. Even in cases when such a balance could be found, its application to estimating the efficiency of using the energy consumed and to analyzing the possibilities for improving a process revealed other drawbacks. In particular, the energy balance drawn up on the basis of the first law of thermodynamics does not allow one to take into account the convertibility of energy forms into work or other forms, in order to determine the locations and causes of all the energy losses.[1]

## 2.2 Complete Energy Balance

### 2.2.1 Derivation

Analysis of technological processes in chemical, metallurgical and other industries uses a "complete" energy balance based on all three laws of thermodynamics [151–168]. The complete balance accounts for all energy forms, including the chemical energy of the fuel, the raw material, products and wastes. In Chap. 1 we introduced the required concepts of chemical energy and exergy of a substance.

[1] This phrase is a contribution to the traditional generally accepted terminology. According to the law of energy conservation, it cannot be lost and can only be converted into other, less useful forms, i.e., it devaluates.

The equation of complete energy balance using chemical energy and exergy but based only on the first law of thermodynamics has the form (a single prime denotes quantities produced, a double prime is for quantities consumed):

$$(\Sigma E)'_{\rm el}+(\Sigma E_{\rm m})'+(\Sigma I_{\rm ch})'+(\Sigma I_q)' = (\Sigma E)''_{\rm el}+(\Sigma E_{\rm m})''+(\Sigma I_{\rm ch})''+(\Sigma I_q)'' \quad (2.3)$$

where $E_{\rm el}$ and $E_{\rm m}$ are the electric and mechanical energy, respectively; $I_{\rm ch}$ is the chemical energy (enthalpy) of fuel, raw material, products and wastes; $I_q$ the thermal enthalpy (heat) brought into the system by raw material, fuel, blast and removed from it by products and wastes, heat supplied to the process or ejected from it by radiation and heat transfer.

Despite the fact that the energy balances given in (2.3) and (2.1) are drawn up only on the basis of the energy conservation law, there are basic differences between them.

The classical (heat) balance takes into account only a part of the available chemical energy, that in the form of heat determined using (2.2) as the result of the chemical reactions. As noted above, the energy balance in the form of (2.3) takes into account chemical energy of all the material flows. It should be noted that the value of the resulting heat effect of the reactions can be determined, if necessary, from (2.3) by using the expression

$$Q_{\rm res} = (\Sigma I_{\rm ch})' - (\Sigma I_{\rm ch})'' \quad (2.4)$$

Studies on real metallurgical processes based on complete energy balance have revealed that $Q_{\rm res}$ found using (2.4) is more accurate than that obtained using (2.2). Thus, although theoretically there should be no difference as to which expression is used, in practice it does exist. Causes of the insufficiently accurate expression (2.2) are given above. Because the values of the specific chemical energy of the final products are relatively small, and even close to zero, errors in these quantities have a minor effect on the accuracy of $Q_{\rm res}$ when using (2.4) and, hence, on the final results of the heat balance. This is confirmed by calculation results carried out in [153].

The energy balance written in the form of (2.3) does not, however, allow the solution of current practical problems on the rational use of energy resources, since it does not take into account the qualitative differences of energy forms used. The introduction of exergy as a performance index for the different energy forms, as it was interpreted by Rant, allowed the equation of complete energy balance of any thermodynamic system to be written

$$\begin{aligned}(\Sigma E_{\rm el})' + (\Sigma E_{\rm m})' + [\Sigma(E_{\rm ch} + B_{\rm ch})]' + \left[\Sigma(E_q + B_q)\right]' \\ = (\Sigma E_{\rm el})'' + (\Sigma E_{\rm m})'' + [\Sigma(E_{\rm ch} + B_{\rm ch})]'' + \left[\Sigma(E_q + B_q)\right]''\end{aligned} \quad (2.5)$$

or, taking into account the fact that the investigator is usually interested in the way the supplied energy devaluates in individual components of the studied system, it can be written as

$$
\begin{aligned}
&(\Sigma E_{el})' + (\Sigma E_m)' + [\Sigma I_{ch}(E_{ch})]' + \left[\Sigma I_q(E_q)\right]' \\
&= (\Sigma E_{el})'' + (\Sigma E_m)'' + [\Sigma I_{ch}(E_{ch})]'' + \left[\Sigma I_q(E_q)\right]'' ,
\end{aligned}
\tag{2.6}
$$

where $E_{el}$ and $E_m$ are again the electric and mechanical energy (identically, exergy); $E_{ch}$, $B_{ch}$ the chemical exergy and anergy of material flows; $E_q$, $B_q$ the exergy and anergy of heat and heat flows. The equation of complete energy balance in the form of (2.6) is simple and convenient for routine calculations and its representation in table form is shown in Table 2.1.

**Table 2.1** Format used to derive complete energy balances

| Balance items | Mass | Energy [kWh] | | Exergy [kWh] | |
|---|---|---|---|---|---|
| | [t] | $I_{ch}$ | $I_q$ | $E_{ch}$ | $E_q$ |
| INPUT | | | | | |
| Raw material | | | | | |
| Auxiliary materials | | | | | |
| Fuel | | | | | |
| Electric energy | | | | | |
| Heat energy | | | | | |
| TOTAL: | | | | | |
| OUTPUT | | | | | |
| End product | | | | | |
| By-products | | | | | |
| Solid waste | | | | | |
| Waste gases | | | | | |
| Energy and exergy losses | | | | | |
| TOTAL: | | | | | |

Isolation of the available part (the exergy) in the energy flows of such a balance allows a more detailed analysis of the total efficiency of using the various energy types as well as a more exact identification of the location and magnitude of the exergy losses at different stages of the process. Apart from the energy efficiency one can also determine the exergy efficiency, which more correctly reflects the thermodynamic goodness of a process.

The energy balance equation can be applied to both operating plants and those still at the design stage as well as to the forecasted technological processes within them. The complete energy balance of an operating plant is written using its material balance, and data on the chemical analysis of all the consumed materials and obtained products and wastes. This allows estimation of the energy use in individual processes of the plant, the value of usable secondary energy resources, their production curves, and operating conditions of the heat recovery installations.

Energy balances for the designed plants and production processes can be drawn up on the base of the past production experience, tests of pilot plants and scientific analysis. These balances are intended for estimating the possible level of energy use of a designed plant as a whole or a new perspective process.

### 2.2.2 Components of the Complete Energy Balance

**Heat and Heat Flow.** In this section we introduce the relations determining the components of the complete energy balance equation discussed above.

The enthalpy and exergy of the heat flow entering the system or leaving it with matter flow, neglecting its kinetic and potential energy, are related according to the expression

$$i_q = i_1 - i_0 = e_q + (s_0 - s_1)T_0 = e_q + b_q \, , \tag{2.7}$$

where $i_1$ and $i_0$ are the specific enthalpies of a substance at the initial parameters and final ones corresponding to the environment; $s_1$ and $s_0$, the specific entropies of a substance at the same parameters; $e_q$ and $b_q$ the specific exergy and anergy of a substance, respectively. Hence, the determination of the energy and exergy in the flow requires knowing the enthalpy and entropy. These data are tabulated [100, 101, 133, 134, 189, 190, 197, 198] or they can be calculated approximately using the formulas given below.

If the heat capacities of the required substances are known in the temperature range of interest and the phase transition temperatures, the enthalpies can be calculated by the expression

$$i_q \cong q = \int_{T_0}^{T_{st}} C'_p dT + \Delta H_{st} + \int_{T_{st}}^{T_{ml}} C''_p dT + \Delta H_{ml} + \int_{T_{ml}}^{T_1} C''' dT \, , \tag{2.8}$$

where $C'_p$, $C''_p$, $C'''_p$ are the heat capacity functions, $\Delta H_{st}$, $\Delta H_{ml}$ the enthalpies of structural change and melting; $T_{st}$, $T_{ml}$ the temperatures of structural change and melting; $T_1$ is the temperature at which the heat content of the substance is determined.

If the above initial data is available, the entropy change during isobaric heating of the substance, in which structural or phase changes may occur, can be found from

$$\begin{aligned} \Delta s = s_1 - s_2 \cong & \int_{T_0}^{T_{st}} (C'_p/T)\, dT + \Delta H_{st}/T_{st} \\ & + \int_{T_{st}}^{T_{ml}} (C''_p/T)\, dT + \Delta H_{ml}/T_{ml} + \int_{T_{ml}}^{T_1} (C'''_p/T)\, dT \, . \end{aligned} \tag{2.9}$$

If the exact dependence of the heat capacity on temperature is not known, one can use the mean heat capacity. In this case (2.5, 6) for the calculation of the enthalpy and entropy change become

$$i_q = i_1 - i_0 \cong q = \bar{C}_p(T_1 - T_0) \tag{2.10}$$

and

$$\Delta s = \bar{C}_p \ln(T_1/T_0) \tag{2.11}$$

where $\bar{C}_p$ is the mean heat capacity of the substance in the desired temperature range.

The exergy of the heat extracted from a hot source with temperature $T_1$ is taken as the maximum useful work due to this heat on the condition, that the environment at the temperature $T_0$ is a cold source. Such work can be supplied in a reversible Carnot cycle. Then the exergy at a constant-temperature hot source equals

$$e_q = q(1 - T_0/T_1) \tag{2.12}$$

and when the temperature varies the exergy is

$$e_q = q - T_0 \int_1^2 dQ/T = q - T_0 \Delta s \tag{2.13}$$

where $\Delta s$ is the decrease in entropy of the hot source during transition from state 1 to state 2.

**Chemical Energy and Exergy of Substances of Given Composition and Structure.** The characteristics of pure chemical compounds and complex mixtures of known composition and structure can be computed only if the specific values of the chemical energy and exergy of all the components are predetermined.[2] Then the chemical energy and exergy of any compound have the form

$$i_z = \sum_i m_j i_j^0 + \Delta H_z^0 , \tag{2.14}$$

$$e_z = \sum_j m_j e_j^0 + \Delta G_z^0 , \tag{2.15}$$

where $i_j^0$ and $e_j^0$ are, respectively, chemical energy and exergy of the element $j$ or elementary substance included in substance $z$; $m_j$is the amount of $j$ in moles; $\Delta H_z^0$ is the standard heat of formation, i.e., the enthalpy change in the formation of substance $z$ from its elementary substances in the standard state; $\Delta G_z^0$ is the standard free enthalpy (the Gibbs energy) of formation, i.e., the change of the

[2] Methods for calculating the chemical energy and exergy of elements and elementary substances are considered in the next chapter.

isobaric potential in the formation of $z$ from its elementary substances in the standard state.

In computing the chemical energy and exergy of complex substances of given composition and structure it is often assumed that without a large error they can be treated as mechanical mixtures. Then using the specific values of chemical energy and exergy of the components of the complex substance, one can calculate the same parameters for the latter by

$$i_{\text{mix}} = \Sigma_z \nu_z i_z \; ; \tag{2.16}$$

$$e_{\text{mix}} = \Sigma_z \nu_z e_z \; , \tag{2.17}$$

where $\nu_z$ is the fraction of the chemical component $z$ in a unit of the considered substance.

Different approaches to calculating the chemical energy and exergy of substances with uniform composition, e.g., solid and liquid fuels as well as different materials such as ores, concentrates, fluxes, slags, mattes, etc. will be considered below.

**Chemical Exergy of Solid and Liquid Fuels.** The role of chemical energy and exergy of fuels in the energy balance of any production process can hardly be overestimated. All forms of fuel belong to a class of so-called conventional energy carriers which play a decisive role in calculating the balances of energy resources of given geopolitical regions.

Fuels are complex chemical substances. The determination of their chemical energy and exergy could be similar to the approach described above, if their composition and chemical formulas were known. The composition of fossil fuels (solid and liquid) is usually non-uniform and unknown. Therefore, $i_z$ and $e_z$ cannot be determined by (2.14, 15), as the values of the standard enthalpy and Gibbs energy of their formation are unknown. Besides, the stoichiometry of the components in the fuel are also unknown. Therefore, the calculation of chemical energy and exergy of fuels with non-uniform and unknown composition requires various approximation methods.

For example, the chemical energy of fuel can be calculated from known characteristics such as high and low calorific values for the fuel mass as-received, moisture-free and dry-and-ash-free ($Q^{\text{rec}}$; $Q^{\text{msf}}$; $Q^{\text{daf}}$) as well as the heat of combustion $\Delta H^0_{\text{com}}$. The heat of combustion and the high calorific value of fuels which are pure organic substances and do not contain sulphur or halogens (S, F, Br, Cl, I) coincide with the value of the chemical energy, i.e., $Q_{\text{h}} = \Delta H^0_{\text{com}} = i_z$. If the fuel contains S, F, Br, Cl, I, which do not form energy-devaluated compounds during combustion, then its chemical energy will differ from the heat of combustion by the amount of the chemical energy of combustion products of sulphur and halogens.

Simplified procedures based on the ratio of the exergy to the high calorific value are also applied for calculating the chemical exergy. Rant proposed the following formulas [131, 132]

– for the chemical exergy of a liquid fuel, whose molecule contains more than one carbon atom

$$e_{ch} = 0.975 Q_h^{rec} \tag{2.18}$$

– for the chemical exergy of a gaseous fuel, whose molecule contains more than one carbon atom

$$e_{ch} = 0.95 Q_h^{rec} \tag{2.19}$$

Rant assumed solid fuel to be a mixture of combustible components C, S, and free hydrogen that would remain if all the oxygen were bound to hydrogen in the form of $H_2O$. This assumption disregards the exergy changes that occur due to the formation of a large number of chemical bonds in the solid fuel which cannot be taken into account exactly. Considering the fact that the chemical exergy of the components C, H and S differs negligibly from their high calorific values, and assuming that the exergy of water contained in the fuel equals zero, Rant proposed that the chemical exergy of hard and brown coals is

$$e_{ch} = Q_{l-el}^{rec} \, . \tag{2.20}$$

Convenient, but more complex formulas for determining the values of chemical exergy of fuels were proposed by *Zakharov* [205]. They take into account qualitative characteristics of fuels such as their ash content ($A^{rec}$), moisture ($W^{rec}$), oxygen content ($O^{rec}$), etc.

Values of the low chemical exergy are estimated using the following equations:

– for hard coals and anthracites

$$e_{ch_l} = \left\{1.009 + (0.131 O^{rec} + 0.116 W^{rec}) / \left[100 - (A^{rec} + W^{rec})\right]\right\} Q_l^{rec} \, ; \tag{2.21}$$

– for brown coals and peat

$$e_{ch_l} = \left\{0.978 + (0.267 O^{rec} + 0.103 W^{rec}) / \left[100 - (A^{rec} + W^{rec})\right]\right\} Q_l^{rec} \, ; \tag{2.22}$$

– for low-sulphur residual oils

$$e_{ch_l} = 1.01 Q_l^{rec} \tag{2.23}$$

– for sulphur residual oils

$$e_{ch_l} = 1.02 Q_l^{rec} \, . \tag{2.24}$$

Primarily hydrocarbon fuels can in many cases be considered, without gross errors, as pure mixtures of carbon and hydrogen. For example, gasoline can be approximated as pure iso-octane, and so on [171].

**Chemical Energy and Exergy of Non-combustibles.** When the energy balances of technological processes are determined, becomes necessary to determine the chemical energy and exergy for a large number of the accompanying non-combustibles, the raw material, products and wastes. As a rule, it is impossible to calculate the chemical energy and exergy for such materials by the same techniques as for uniform substances. This is because for them, as well as for non-uniform substances, it is not always possible to exactly determine the chemical bonds between the components and individual groups. Chemical analysis usually reveals the content of only the most important components and individual chemical groups (CaO, MgO, $SiO_2$, FeO, etc.) but it does not indicate into which compounds they are bound. Thus, such a description of the composition is insufficient for determining the chemical energy and exergy, since the chemical energy (exergy) of a compound is not the sum of the chemical energies (exergies) of its groups. Hence, chemical analysis alone is insufficient and must be supplemented by a knowledge of the mineralogical structure.

The mineralogical composition can be determined by examination with a microscope of polished samples, or other methods of analytical chemistry. This analysis allows one to compile the data of the most frequently occurring minerals, which along with the data of the chemical analysis, can be used for calculating the so-called rational mineralogical composition. The rational composition can be the basis of choosing and optimizing a metallurgical processing method.

Such data on the mineralogical composition of various raw materials, intermediate technological products and wastes are available. The specific values of chemical energy and exergy can thus be determined using (2.16, 17).

In the analysis of complete energy balances it is advisable to preliminarily compute the composition and specific values of chemical energy and exergy of all the initial and final non-combustibles. Although this work is sufficiently laborious, it is compensated in the long run by considerable simplification.

## 2.3 Solving Practical Problems

In our opinion, the most important practical problems which can be examined using energy balance methods are

- determination of the efficiency of energy use;
- identification of the places and causes of energy (exergy) losses;
- development of measures for reducing the energy consumption and losses;
- determination of limiting indices on energy consumption and conservation.

It should be emphasized that the author adheres to the opinion that this analysis does require the complete energy balance to be drawn up for the following reasons. Many experts, including *Szargut* [173–176, 184], who was the first to propose the concept and method for calculating the chemical energy of substances, gave up the analysis of energy flows and emphasized the studies on exergy flows and exergy balances. Taking into account that exergy does not follow the conservation law, they introduced the values of exergy losses caused by irreversibility into the output part of the balance. These losses can be determined by using the Gouy-Stodola law (1.53).

Therefore, to make the exergy balance it is necessary to perform a detailed entropy analysis for such aggregate systems. However, studies using the complete energy balance written in the form of (2.6) allow one to avoid this laborious procedure in energy problems that do not require the detailed analysis of these losses. The total losses due to irreversibility are then the difference between the exergy flows in the input and output parts of the complete energy balance.

The complete energy balance presents the picture of the distribution of all energy (exergy) forms in the energy consumers, and their actual use by these consumers, with account of transport losses of energy resources, raw material, products and wastes. Such a picture closely relates the production scheme (material flows) and the energy (exergy) flows of all energy forms. Therefore, it can serve as a basis for improving both the individual parts of a production process as well as the entire production plant. This information is useful for experts dealing with operational problems and for those who designs new objects. It can prompt them new versions of the equipment arrangement that may lead to the reduction in energy losses for transport of energy resources and other materials, promote the development of measures for increasing the efficiency of energy use in the production process as a whole due to rational combination of energy flows in its constituents.

### 2.3.1 Determination of Energy Use Efficiency in a Process. Idealized and Ideal Analogs of Processes

The determination of efficiency of energy use is one of the more important problems that can be solved using the energy balance method.

A technological process, workshop, plant, industry, as well as any thermodynamic system can be characterized by an energy efficiency, which is determined from their energy balance. Assignment of the value of chemical energy to all substances participating in the thermodynamic process makes the calculation of its energy efficiency rather simple and sufficiently definitive, unlike the determination of the thermal efficiency.

Both the energy $\eta_{en}$ and exergy $\eta_{ex}$ efficiencies can be determined. These efficiencies are simply the ratios of the "useful effect" of interest to the total energy or exergy consumption, i.e.,

$$\eta_{en} = \Sigma I_{us} / \Sigma I_{cons} \tag{2.25}$$

$$\eta_{ex} = \Sigma E_{us} / \Sigma I_{cons} \,. \tag{2.26}$$

In calculating the energy efficiency, usually the different degrees of usefulness of energy forms is not taken into account. Strictly speaking, this efficiency index should be used only in estimating the efficiency of processes where entropy-free energy is consumed and produced as the useful effect. Exergy efficiency does not have this restriction but, unfortunately, at present it is not commonly used.

The values of these efficiencies reflect the thermodynamic goodness of any process. In some processes where the useful effect and energy consumption are expressed through entropy-free energy or work, the energy and exergy efficiencies can be the same. If the energies of different degrees of usefulness are utilized, the thermodynamic efficiency is more correctly characterized by the exergy efficiency.

Both energy and exergy efficiencies can be determined rather simply and definitively if the "useful product" is well defined. Then energy and exergy flows at the process input represent the energy consumption and the corresponding flows at its output with useful products are the useful effect.

In our view, useful products should include both the desired products of the process and partly finished products that are transferred to another stage in the production scheme for further processing. In this case the energy and exergy efficiency become intimately related to the particular technological processes and the efficiency of multi-product processing of the raw material. The more complex the process, the larger the amount of useful products obtained and the higher the energy and exergy efficiency.

The absolute values of energy and exergy efficiency determine from (2.25, 26) yield only the values of the achieved level of energy use in the studied process, but do not show the limits on its further rise. The goodness of a process can be revealed only by a comparison of the efficiency values with the currently maximum and theoretically feasible ones for the given process.

In practice, the efficiency of a new production process is sometimes estimated by comparing its techno-economic indices to those of the best existing technologies. The disadvantage of this approach is that the state of the existing technologies is continuously progressing, so that this cannot be a constant basis for comparison. Therefore, it seems more reasonable that efficiency estimations should be based on a comparison to theoretical limiting thermodynamic indices for an idealized and ideal analogs of the process. The theoretical indices independent of any transient factors.

The concept of an idealized and ideal analog in estimating the energy efficiency of processes and machines has been used in engineering science for many years. Depending on the nature of the problems, analogs with different degrees of idealization are used. For example, along with the Carnot cycle, which is a function of only the temperature of the hot and cold heat sources (maximum possible idealization), Rankine and Diesel cycles, among others, which take into account the physical properties of the working body are used in thermal power engineering. In the following discussion we show how to devise an idealized analog for the physico-chemical transformations that are the particularly emphasized subject of this monograph [152, 154, 157, 168].

The process whose exergy efficiency equals unity is defined as the analog of a technological process with the maximum degree of idealization (or an ideal analog). By definition, the chemical exergy of a substance characterizes the minimum work required for its production from the natural substance using the corresponding reversible reaction (reaction of devaluation). Hence, processes described by the corresponding reactions of devaluation can be taken as the ideal analogs of industrial chemical transformation processes. The values of the chemical energy and exergy of a product thus give the absolute minimum energy consumption and amount of work required for its production by any possible technology.

For analogs with a lesser degree of idealization, we assume that any real process can be described completely by a sequence of reactions (or by a reaction). These reaction steps require stoichiometric amounts of the reagents and proceed to the end. Material and energy exchange at the in- and outputs of the process are determined by the parameters of the environment. The amount of energy consumed in the transport of the initial reagents and final products is equal to zero. With these assumptions analogs of real production processes can be obtained by combining the idealized analogs of individual processes as in the given production scheme.

The complete energy balance can also be applied for study of the efficiency of idealized processes. The complete energy balance of the idealized analog is used to calculate the energy $\eta_{\text{en}}^{\text{id}}$ and exergy $\eta_{\text{ex}}^{\text{id}}$ efficiency using expressions similar to those for determining these characteristics of the real processes. The values of these efficiencies are estimates of the maximum possible values for the corresponding real process. The ratios of these quantities serve as indicators of the degree of thermodynamic ideality, i.e.,

$$\eta_{\text{en}}^{\text{rel}} = \eta_{\text{en}}^{\text{real}} / \eta_{\text{en}}^{\text{id}} \tag{2.27}$$

$$\eta_{\text{ex}}^{\text{rel}} = \eta_{\text{ex}}^{\text{real}} / \eta_{\text{ex}}^{\text{id}} \ , \tag{2.28}$$

The higher the values of $\eta^{\text{rel}}$, the higher the energy effectiveness (and the fewer the possibilities of further improvement!).

### 2.3.2 Energy Losses and Secondary Energy Resources

In this section the means of finding the distribution of energy consumed in a multi-step, multi-product process is described. It requires as input, a complete energy balance, which we have shown to be derivable from the exergy method. In addition to energy consumed directly, estimates of the amount of energy lost, and the amount that might be recovered are recessing.

The exergy method allows the correct estimation of losses found from the energy (heat) balance. The nature of these losses, and their impact on the economic efficiency, are not always defined in the same way. The exergy method reveals, in particular, losses that cannot be estimated by using the first law of thermodynamics, but do notably effect the thermodynamic efficiency, for example, losses due to irreversibility of chemical reactions, as in combustion, losses due to non-equilibrium heat exchange, losses due to mixing, throttling, etc. Exergy losses due to irreversibility are unrecoverable. Energy losses which are recoverable become secondary energy resources (low-temperature heat) and can also be found from a complete energy balance analysis. (The determination of the amount of secondary energy resources, in operating and planned plants, which can actually be included into energy budgets of regions or country as a whole in the current and long-term planning is among the most important problems which are treated with the energy balance method.) It is necessary to know not only their absolute value at a given time, but also their time dependence, that is, the rate of secondary energy resource production and consumption. As discussed previously, the exergy value characterizing the ability of a given kind of energy to do work or to transform it into some other form, can be a necessary and sufficient characteristic of the value of resources theoretically suitable for further use. For example, the chemical exergy of wastes characterizes their potential for further use as raw material or as an intermediate product for further processing, or as secondary fuel energy resources. Physical exergy value in the wastes of some or other process indicates the possibility for their use as thermal secondary energy resources.

However, the complete energy balance determines only the "theoretically suitable" secondary energy resources. In actual applications it is necessary to determine the technically suitable and then economically expedient secondary energy resources. These problems are considered in greater detail in Chap. 4.

Accurate, scientifically based standardization of the consumption of energy resources is of crucial importance. It not only promotes technological progress but also increases labor productivity. Development of standards on the specific energy consumption per unit of product or labor is a necessary base for calculating the demands of a process, a workshop, or a plant for energy carriers. These data, in turn, are the input data in budgeting the demand for energy resources of economic sectors, which finally, allows a nation's energy balance to be derived.

Complete energy balance is the most appropriate means of standardization of the consumption of different energy resources. It can be the basis for determining the specific consumption of conventional energy resources with account of the chemical energy (exergy) of raw and auxiliary materials entering the process. In addition, isolation of the exergy fraction in energy flows allows the consumption of energy for a unit product output to be determined. Another advantage of the complete energy balance is that its data allow a better defined distribution of energy used in multi-product processes and production schemes.

Among the methods proposed for the distribution of expenditures, particularly energy expenditures, the exergy method seems to be the most convenient and valid. Its essence consists in the following [33, 53, 57, 82, 116, 117]. We assume that the process under examination has consumed the energy $I_{cons}$ and yields $n$ types of product $i$, thus, $I_{cons} = \sum_{i=1}^{n} I_i$. The amount of energy carrier required for the product $l$ is proportional to the fraction of its exergy in the total exergy of all products

$$I_l = I_{cons} \left( E_l / \Sigma_{i=1}^{n} E_i \right) . \tag{2.29}$$

Energy losses are divided in proportion to the fraction of useful energy.

Usually the operation of a plant has a branched scheme: each product is an input for a subsequent process. Therefore, the energy consumed for a product $i$ in a process $j$ should be included into the energy consumed by the process $(j + 1)$, where this product is further processed. Then the energy consumed by process

$(j+1)$ is the sum of $I_i$ and the energy consumed directly in this process, and the total energy consumption for each final product $l$ is

$$I_l = \left\{ \left[ I_1 \left( E_{l_1} / \Sigma_i E_{i_1} \right) + I_2 \right] \left( E_{l_2} / \Sigma_i E_{i_2} \right) + \ldots + I_r \right\} \times \left( E_{l_r} / \Sigma_i E_{i_r} \right) \tag{2.30}$$

where $E_{l_1}$ is the exergy of the product due to the first process, $j = 1$; $E_{l_2}$ is the same for the second process $j = 2$; $E_{l_r}$ is the exergy of the final product $l$ due to process $r$; $I_1, I_2, \ldots, I_r$ are the energies consumed in the first, second, … $r$-th processes, respectively.

Application of the exergy-based criterion for distribution of consumed energy is the most appropriate for processes involving complex energy and physico-chemical transformations. One only needs to take the preliminary step of calculating the complete energy balance to find both the energy and exergy values.

## 2.4 Theoretical Potential and Energy Reserves

From the moment of its initial startup, every process is continually improved and reoptimized for better energy and exergy efficiency. The efficiency can be seen to change in time in an ascending curve which asymptotically approaches the limiting, ideal efficiency of the given process. Hence, the energy consumption also tends to some limit, the lowest, practically unachievable value, i.e., to that of an idealized or ideal analog.

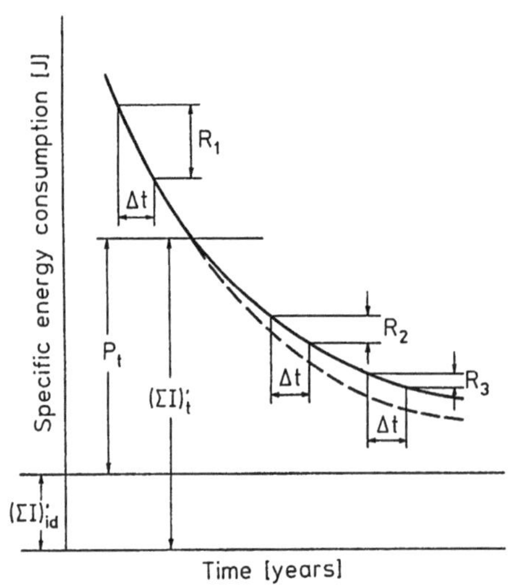


**Fig. 2.1.** Determination of potential and reserves of energy conserved due to process improvement. The two curves show the effect of a more efficient energy conservation program (dashed line)

Figure 2.1 shows the energy saving potential $P_t$ at the time $t$ which is the theoretically feasible amount by which the energy consumption can be decreased to the energy consumption level in the idealized or ideal analog $(\Sigma\Gamma)'_{\mathrm{id}}$. In other words, the energy saving potential is the maximum possible difference between the achived $(\Sigma\Gamma)'_t$ and theoretical energy (exergy) consumption. Similarly to the actual energy consumption, which decreases over time due to process optimization towards the energy consumption value of the idealized or ideal analog, the energy saving potential asymptotically approaches zero with increasing time.

The total energy conserved is the difference between the energy consumption of the process or production scheme at the base and future levels:

$$R_t = (\Sigma\Gamma)'_t - (\Sigma\Gamma)'_{t+\Delta t} \tag{2.31}$$

The energy saving potential at a time $t + \Delta t$ depends to a great extent on the energy saving potential at time $t$, i.e., on the point of the curve that corresponds to the current state of technology. Thus, when a process has already been extensively optimized the possibilities for decreasing the energy consumption further, the energy saving potential can be several times lower during the same time interval $\Delta t$ where the slope of the curve is low, than for the initial part of the curve, i.e., in Fig. 2.1: $R_3 < R_2 < R_1$. This means that when estimating the energy saving potential it is necessary to correctly determine the current stage of development of the process, that is, which part of the curve corresponds to the current state-of-the-art. It is also necessary to correctly relate the achieved level of energy consumption to the minimum possible one and know the degree to which the former approaches the latter. It is the introduction of technology analogs with different degree of idealization that favours such analysis.

Thus the energy reserves due to improvements can also be determined from the difference between the current energy saving potential and the prospective level:

$$R_t = P_t - P_{t+\Delta t} \,. \tag{2.32}$$

Since $P$ tends to zero, $R_t$ also tends to zero.

The time dependence of energy consumption may change depending on the effectiveness of the energy conservation policy. With a more effective policy the slope of the curve becomes more negative (dashed line in Fig. 2.1).

The development in an industrial sector depends on improvements in the composition of production processes. Obsolete, inefficient equipment and technologies are replaced by ones with better techno-economic and energy indices. This process causes changes in the overall production structure of an industry. The determination of the energy saving potential and energy reserves due to retirement of obsolete technologies and introduction of new ones is somewhat different from the one given above and is presented in Fig. 2.2. It gives the characteristics of the time-dependent energy consumption change due to the removed components ("technology 1") and those introduced ("technology 2").

If technology 1 is substituted by a new technology 2 at $t_0$ (base level), then the energy saving resources at the future level $t_1$ can be determined from

$$R_{21} = (\Sigma\Gamma)'_{1_{t_0}} - (\Sigma\Gamma)'_{2_{t_1}} \tag{2.33}$$

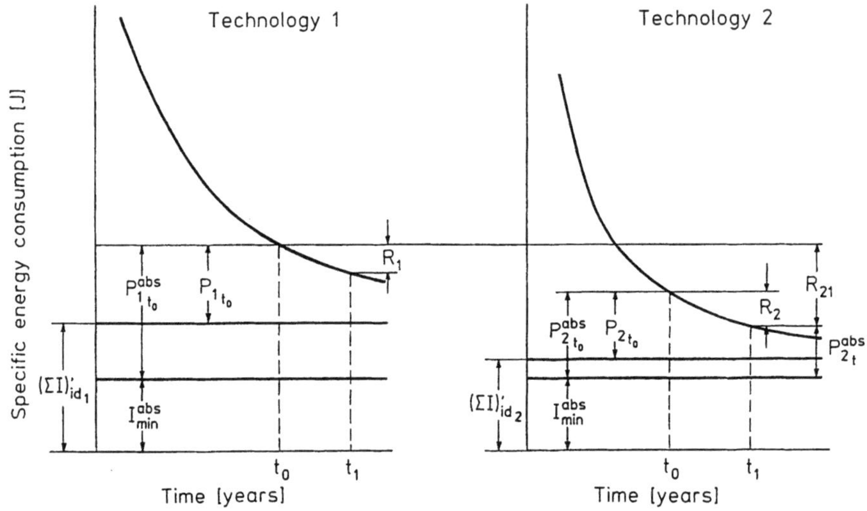


**Fig. 2.2.** Determination of potential and reserves of energy conserved due to substitution of current technology by new, more efficient methods

where $(\Sigma I)'_{2_{t_0}}$ is the specific energy consumption using technology 1 at the start of the period and $(\Sigma I)'_{2_{t_1}}$ is the same for technology 2 at the end of the considered period. Thus, these calculations take into account the possibility for further improvement of technology 2 during the period between $t_0$ and $t_1$.

The idealized analogs of the technologies under consideration very often do not coincide; as seen in the figure. Then the estimation of the energy saving potential due to technology substitution is possible only by determining the absolute minimum energy consumption for any technology producing a product from the given raw material. This is the definition of the energy consumption of an ideal analog with the maximum degree of idealization of manufacturing this product. An example of such an analog is the Carnot engine, whose operating cycle consists of only reversible processes and its exergy efficiency equals unity.

As was noted above, the chemical exergy of a substance, by definition characterizes the minimum work done to produce the substance from some natural substance by using the appropriate reversible reaction (reaction of devaluation). The process described by this reversible reaction or by several reversible reaction steps, is the ideal analog of real production processes based on chemical transformations. For this analog, the values of chemical energy (exergy) of the considered substance characterize the absolutely minimum energy and work needed for its production by any means.

Therefore, if the absolute minimum energy consumption $I^{abs}_{min}$ (the lower line in Fig. 2.2) which is the lower limit for any existing or future technologies of manufacturing product $i$ can be determined, these characteristics can be applied to find the absolute energy saving potential $P^{abs}$ for each considered technology. Then we can find the energy saving reserves due to the transition from one technology to another:

$$R_{21} = P^{\mathrm{abs}}_{1_{t_0}} - P^{\mathrm{abs}}_{2_{t_1}} \tag{2.34}$$

To find the energy saving potentials due to process improvement, analogs with a lower degree of idealization, for example Rankine or Diesel cycles which take into account properties of the working body should be used. To study the energy saving potentials in a transition to more efficient technologies, the absolute minimum of energy needed for the product output should be determined using the analog with the maximum possible degree of idealization.

Idealized and ideal analogs which allow the determination of theoretical energy saving potentials, enable a comprehensive analysis of optimization programs. They also allow one to evaluate, how close the limiting levels have already been approached. With decreasing $P_t$, increase in the process efficiency requires ever more complex and expensive measures.

The concept of ideal and idealized analogs for production processes are quite useful in the standardization of energy consumption and prohibits overly optimistic prognoses when new plants or new processes are brought on line.

In conclusion, it should be noted that this discussion is valid only for the estimation of the amount of energy conserved which can be realized technically and does not include the economic aspects of the problem. Problems of economic efficiency will be briefly considered later.

# 3. Calculation of Chemical Energy and Exergy of Elements and Elementary Substances

The chemical energy and exergy of substance are calculated from the values of these quantities for the elements composing the substance. The correct determination of the latter to a great extent specifies the accuracy of calculations and resulting energy and exergy input of a considered process, in particular, with non-conventional energy carriers (processed raw material, fluxes, auxiliary materials, etc.). In this chapter the methods for determining the chemical energy and exergy proposed by different authors are renewed and compared.

## 3.1 Choice of the Environment Model

The choice of a reference level determines the absolute values of chemical energy and exergy of substances and thus other energy characteristics. This reference level, defined by the parameters of the environment therefore determines the method of realization and efficiency of production processes. Whether the considered system is closed or open, its operation is accompanied by either exchange of only energy or energy and matter with different environmental components characterized by a set of parameters.

In the exergy method of analyzing processes the role of the environment is critical. From the definition of exergy it follows that a set of environmental parameters must be defined as a reference level. In real systems the parameters of the environment usually vary in space and time over a non-negligibly wide range. Thus, it is necessary to devise a model of an idealized environment which, however, does approximate a real system.

The selection of such a model is rather complex and not unambiguous and has been considered in a large number of publications. It should be emphasized that the selection of the model for the environment and the method for calculating the chemical exergy are closely interrelated.

The model of the environment must meet two conditions: The first condition is that the parameters of the environment must remain constant during any interaction with the system. Alternatively, the environment must be so large with respect to the system that an interaction would not lead to a significant change in the parameters of the environment.

Thus, in analyses of real system it is necessary to take marked changes in environmental parameters, especially within some localized region in the direct vicinity of the systems into account. An increase in the capacity of technical systems and a localization of industrial production will result in a significant change in the environmental parameters.

The second condition is that all the components of the environment should be in complete thermodynamic equilibrium. Only in this case can it be characterized by zero exergy and therefore can be used to define a reference level. Note that the natural environment is not and cannot be in complete equilibrium, primarily because of an ever-evolving biosphere and of course, cosmic and solar radiation.

Although these conditions are understood and accepted universally, much dispute remains as to the correctness of the environmental model. Szargut, the originator of the concepts of chemical energy and exergy of substance suggested using a reference system based on the following principles [174–176, 180, 181]:

1. The environmental model includes only the atmosphere and the upper layer of the lithosphere, but without any indication of the layer thickness. Later components of the hydrosphere, ions and neutral substances contained in sea water were added.

2. It is assumed that at the standard values of pressure and temperature ($p_0$ and $T_0$) all the components of this environment are in thermodynamic equilibrium and hence their chemical energy and exergy are equal to zero.

3. The most widely spread (and as a rule mst devaluated) substance within the corresponding bounds of the real environment is recommended to be selected as a reference species for each element.

Ahrendts raised questions about both Szargut's technique and model of the environment [2, 3]. He performed detailed calculations of compositions of the environment that were at thermodynamic equilibrium at standard conditions. His environment comprised components of the atmosphere, sea water and an external stratum of the earth's crust. The calculations led to the following conclusions:

– the equilibrium composition of the environment substantially differs from the real one; the equilibrium atmosphere should contain oxygen in very small amounts; nitrates of Na, K, Ca replace oxygen

– the equilibrium composition depends to a great extent on thickness of the earth crust layer, whose components are included in the calculation.

Ahrendts showed that the composition of the environment determined by the equilibrium of the atmosphere, ocean and a stratum of the earth crust 1 m thick was practically identical to the composition of the real environment if judged on the composition of the atmosphere. He argued that the model of such an environment, taken as a reference level, could have satisfied thermodynamic requirements most accurately. However, from the practical viewpoint such a model is not suitable since it does not contain many elements that are of interest, for us, in particular for the metallurgical processes.

The environmental model proposed by Szargut is further criticized in [36, 150]. These authors discovered a shortcoming in that only one precisely defined reference species corresponds to each considered substance independent of the conditions of solving problem. They think it paradoxical that in obtaining an element from different reference species the same value of exergy is assigned to this element, although in each case the minimum work required to extract it is different. On this basis, the authors conclude that the chemical exergy of substances should be calculated in conformity with the conditions of the extraction process. In addition to the two conditions for the environmental model mentioned above they formulate one more; the exergy of a substance calculated by this model must not significantly differ from the minimum work required to extract and process it under the real conditions of the system under analysis. If this is not the case, the authors of [36, 150] suggest to choose unique reference system for each technical system.

It is difficult to agree with this conclusion, since such an approach would rule out the possibility to compare even different apparatuses of a monotype of single type technical systems with each other. It would result in still greater paradoxes, in particular, when the product of one process is used as a raw material in another located elsewhere, under different conditions. Such an approach to the choice of environmental model completely excludes the possibility of standardizing the chemical exergy values for different elements and their compounds. This standardization is one of the most important advantages of the Szargut technique as well as similar models [75–78, 112–115, 120–122, 151–168, 172–187, 201]. The universal environmental model allows the calculation of standard values of the chemical energy and exergy of substances, and use them as reference data.

Many energy problems do not require taking into account the deviation of real environmental parameters from standard values. At the same time it is necessary to clearly understand that there are such problems whose solution using the standard values of energy and exergy is not correct. It is precisely such problems that are considered in [36, 150] and the methods of their solution using the models of a local environment, or a controlled environment are justified.

## 3.2 Short Overview of Methods

Szargut introduced the concept of chemical enthalpy (energy) and exergy of the substance and suggested the technique to determine these values for chemical elements [180, 181]. A reference species, i.e., completely energy devaluated substance, under standard conditions in the environment, is established for each element. Then the devaluation reaction, i.e., the reaction for an idealized process transforming the element to its reference species, is derived for each element. In this reaction only the element under consideration and the species of its chemical elements are used.

The normal chemical exergy, according to Szargut, is the sum of the maximum work in the devaluation reaction which can be directly calculated as the sum of the Gibbs energy change in this reaction and of another component determined by concentration of reference species in the environment. In calculating this component it is assumed that all the reference species in the environment behave as components of an ideal solution. As always, the relative concentration of a gaseous reference species in the atmosphere corresponds to its partial pressure in air. The concentration of solid reference species in the lithosphere is determined from geochemical data (Clark, Goldschmidt) on the average abundance of elements and compounds in the earth's crust.

Szargut recognized that the assumption in the chemical exergy calculation of solid reference species acting as components of an ideal solution is an approximation, which can affect the magnitude of the predicted. For this reason, later he modified his initial method of finding the reference species by including the ionic and neutral components of sea water [182].

In [112–115, 173, 178, 183, 184] the formulation of the method of calculation of the chemical exergy of elements was further developed. The reference species were chosen to be components of the atmosphere ($O_2$, $N_2$, $CO_2$, $H_2O$, $D_2O$, Ar, He, Ne, Kr, Xe), ions and non-ionized substances contained in sea water and the solid reference species from the lithosphere. It should be emphasized, however, that if solid substances are used as reference species, the chemical exergy can be estimated only approximately. The procedure of calculating the standard concentration of reference species in the lithosphere is in principle wrong. In addition, it should be noted that in many cases when determining the ion activity factors obtained values seem unreliable. In any case, it has been shown that their value has only a small influence on the chemical exergy calculations. The advisability of including the concentration component into the chemical exergies, determined under such assumptions and to such an accuracy, is not discussed by the authors.

The most important conclusion made in [115, 183, 184] is that to make the exergy calculations easier it is absolutely necessary to define standard chemical exergies whose values for different elements and compounds can be calculated and tabulated. Such data would greatly facilitate the solution of a large number of general energy problems of the higher hierarchical levels than the level of an individual process or aggregate. In [115, 183, 184] it is noted that deviation of the real environmental parameters from the reference values should be taken into account only when it is really necessary, i.e., when neglect of the deviation can significantly influence the results of the calculation.

Yamauchi and Fueki introduced new thermodynamic functions (the theta function and the reference exergy), which actually are exactly the same as the chemical energy and exergy of elements [201]. These authors use a calculational scheme and initial assumptions, that differ negligibly from those proposed by Szargut. Reference species are chosen from the most widely spread compounds present in the environment (the atmosphere and lithosphere) and are so stable that no energy can be produced from them by chemical reaction.

The calculation of the chemical energy and exergy is carried out in a sequence of steps. First, all the elements are arranged. The elements O, N, C and H occupy the first four places, the rest are arranged according to their Clark units. The chemical energy of air, oxygen and nitrogen are set equal to zero. The chemical exergy of O, N, He, Ne and Ar are determined with respect to their concentration in the atmosphere, whose chemical exergy is assumed to be zero. For example, the chemical energy of carbon is determined by enthalpy of formation of carbon dioxide and the chemical exergy, that is, by using the Gibbs energy of formation of $CO_2$ and its partial pressure in the atmosphere.

The chemical energy and exergy of hydrogen is determined by the enthalpy and Gibbs energy of formation of liquid water, respectively. The energy and exergy of atomic oxygen, nitrogen and hydrogen are taken to be the values obtained for 0.5 mole of $O_2$, $N_2$, and $H_2$.

Then the reference species are established and the chemical energy and exergy of other elements are determined. The reference species of element $M$ is selected among its compounds with other elements of lower ordinal number than for $M$.

It follows that the reference species $MX_x$ must be chosen so that it should contain element $X$, whose chemical energy and exergy are known. Then these characteristics for element $M$ can readily be determined by the values of formation enthalpy and Gibbs energy of reference species.

In general, elements have the same reference species for both the chemical energy and exergy. In some cases, however, different reference species must be used because of the lack of a complete set of dynamic quantities.

The given ordered technique allows the values of chemical energy and exergy for the majority of elements used in practice to be determined.

Many other methods for calculating the chemical energy and exergy of substances have recently been published in addition to those mentioned above [25, 64, 83, 84, 109, 110, 143, 144, 172]. In the next subsection we will be discussing one of them in more detail.

### 3.2.1 The Simplified Ozoling-Stepanov Technique

In the model developed by *Ozoling* and *Stepanov* [120, 152, 153], in particular the parameters of the environment were reviewed and, independently of Szargut and co-authors, the hydrosphere was included as a main component. A second goal was the simplification of calculating the chemical energy and exergy. The introduction of the hydrosphere into the model of the environment allows the extension of the number of reference species and the use of a broader range of thermodynamic quantities.

These quantities may include the heat and Gibbs energy of forming aqueous solutions, the enthalpy and free enthalpy of forming ions and, finally, standard electrode potentials. The addition of these parameters enable, in our opinion, a rather simple technique to be obtained, which we present below.

Taking infinitely dilute aqueous solutions as the zero level of the chemical energy and exergy, the values of enthalpy and free enthalpy of devaluation of elements forming water-soluble compounds which were originally determined by Szargut can be made more accurate. This is done by taking into account the changes in the enthalpy and entropy in the formation of infinitely dilute aqueous solutions of the appropriate reference species. Such data have been extensively tabulated.

The formula for the chemical energy of an element (elementary substance) is

$$i_{\rm elm} = \Sigma_j \left(\Delta H^0 + \Delta H_{\rm dis}\right)_j - \Sigma_k \left(\Delta H^0 + \Delta H_{\rm dis}\right)_k \tag{3.1}$$

where $\Delta H_{\rm dis}$ is the heat of solution, i.e., the enthalpy change required to form an infinitely dilute water solution of the reference species. The entropy change in the devaluation reaction is then

$$\sigma_{\rm elm} = S^0_{\rm elm} + \Sigma_j \left(S^0 + \Delta S_{\rm dis}\right)_j - \Sigma_k \left(S^0 + \Delta S_{\rm dis}\right)_k \ , \tag{3.2}$$

where $\Delta S_{\rm dis}$ is the entropy change in forming the infinitely dilute water solution of the reference species. The use of (3.1, 2) is restricted due to the insufficient quantity of reference data on enthalpy and entropy changes of formation of aqueous solutions. However, other quantities which can also be used to determine the chemical energy and exergy of substances, namely, the enthalpy and Gibbs energy of formation of a substance in aqueous solution [65, 85], are more readily available.

Since we have taken the infinitely dilute solutions of substances in water as the reference level it is assumed that their chemical energy and exergy are equal to zero. Then using the law of energy conservation we may write the following equalities:

$$\Sigma_j m_j i_j^0 = -\left(\Delta H_z^0\right)_{\rm aq} \tag{3.3}$$

$$\Sigma_j m_j e_j^0 = -\left(\Delta G_z^0\right)_{\rm aq} \ , \tag{3.4}$$

where $(\Delta H_z^0)_{\rm aq}$ and $(\Delta G_z^0)_{\rm aq}$ are the enthalpy and free enthalpy (the Gibbs energy) of formation of a substance in aqueous solution, respectively.

Relations (3.3, 4) allow one to determine the values of the chemical energy and exergy of an element when these values are known for all the other elements of the considered substance.

Standard electrode potentials of elements in aqueous solutions, collected in e.g. [43, 126, 128], can also be used to calculate the chemical energy and exergy of substances. These potentials are determined experimentally with respect to some reference standard, usually a hydrogen electrode under standard conditions.

The electromotive force of the reversible galvanic element and the heat evlved in the net reaction are related via the Helmholtz equation

$$\varepsilon = \frac{\left(\Delta H_z^0\right)_{\rm aq}}{n_{\rm e} F} + T\left(\frac{\partial \varepsilon}{\partial T}\right)_p \tag{3.5}$$

where $F$ is the Faraday constant $\sim 96.5$ kC/mole; $n_{\rm e}$ is the number of electrons exchanged in the reaction; $\varepsilon$ is the electromotive force of an element the difference between the anode and cathode potentials. From (3.5) and the Gibbs-Helmholtz equation for isobaric processes we have

$$G = H - TS = H + T\left(\frac{\partial G}{\partial T}\right)_p$$

where $H$ and $G$ are the system enthalpy and free enthalpy as always $S$ is the entropy. It follows that the electromotive force of an element and the potentials of the electrodes are related to the free enthalpy of the reaction:

$$\varepsilon = \varphi_c^0 - \varphi_a^0 = \left(\Delta G_z^0\right)_{aq}/n_e F \, , \tag{3.6}$$

where $\varphi_c^0$ and $\varphi_a^0$ are the cathode and anode potentials, respectively. Since the free enthalpy of the total reaction is given by the Gibbs energy values of the individual reaction steps, relation (3.6) can be written

$$\varepsilon = \varphi_c^0 - \varphi_a^0 = \left(\Delta G_z^0\right)_{aq}/n_e F = \left(\Delta G_c^0\right)_z/n_e F - \left(\Delta G_a^0\right)/n_e F \tag{3.7}$$

where $\Delta G_c^0$ and $\Delta G_a^0$ are the changes of the isobaric-isothermic potentials (the Gibbs energy) in forming the cation and anion from the elementary substances in aqueous solution.

Thus, the values of chemical energy and exergy of elements can be determined from (3.6,7). However, note that it is necessary to take into account that the tabulated values of electrode potentials as well as the values of the enthalpy and free enthalpy of ion formation must be corrected due to the different reference level. We shall illustrate this on the basis of the electrochemical standard potentials.

As mentioned, the hydrogen electrode is usually taken as the reference standard. The reaction which occurs on it is $2H^+ + 2e^- \rightleftarrows H_2$. The isobaric-isothermic potential (the Gibbs energy); i.e., the standard electrode potential of this reaction, is set equal to zero. In our technique, however, the chemical energy and exergy of water and oxygen, rather than hydrogen potentials, are taken as the zero level. Thus, to determine the potential of hydrogen electrode in our reference scheme we write the net reaction proceeding in the ideal fuel element,

$$H_2 + 0.5O_2 \rightleftharpoons 2H^+ + O^{2-} \rightleftharpoons H_2O \, .$$

The Gibbs energy of this reaction is $\Delta G^0_{H_2O} = 237.34$ kJ/mole. Then according to (3.7) the difference of the electrode potentials of this reaction is equal to

$$\varepsilon = \Delta G^0_{H_2O}/n_e F = -237/2 \cdot 96.5 = -1.23 \text{ V} \, ,$$

which is the resulting potential of a hydrogen electrode. The Gibbs energy and enthalpy of water formation are the hydrogen exergy and energy, respectively, i.e.,

$$e^0_{H_2} = -\left(\Delta G^0_{H_2O}\right) \quad \text{and} \quad i^0_{H_2} = -\left(\Delta G^0_{H_2O}\right) \, .$$

The essence of the procedure of changing the reference level is illustrated in Fig. 3.1. It shows the potentials of some metals relative to the zero (standard) potential of reaction of hydrogen ion formation. After the reference level is changed the hydrogen ion potential is equal to 1.23 V. This means that the new zero level is higher than the standard one by the indicated value that is shown in the figure. The values of the electrode potentials recalculated for the new reference level are

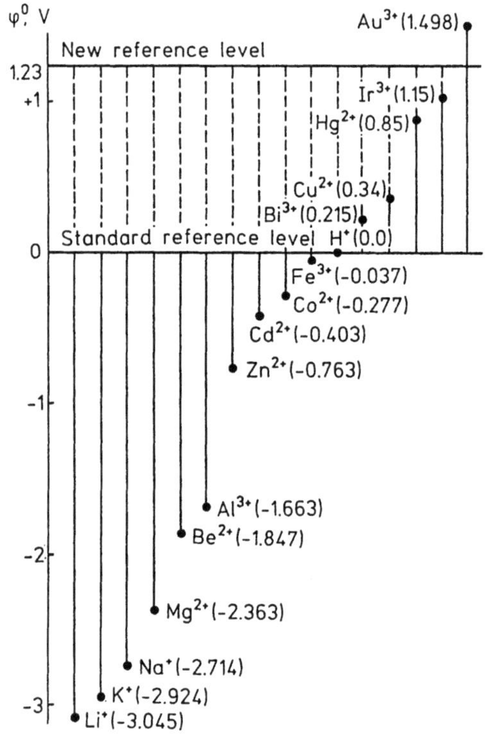


**Fig. 3.1.** Standard electrode potentials

$$\varphi_{\text{ion}}^{*} = \varphi_{\text{ion}}^{0} + (-1.23) \, , \tag{3.8}$$

where $\varphi_{\text{ion}}^{*}$ is the electrode potential of the ion determined relative to the accepted reference level. Then the Gibbs energy value of ion formation in aqueous solution is found from

$$\left(\Delta G_{\text{ion}}^{0}\right)^{*} = \pm \left[\varphi_{\text{ion}}^{0} + (-1.23)\right] n_{\text{e}} F = \pm \left(\varphi_{\text{ion}}^{*} n_{\text{e}} F\right) \, . \tag{3.9}$$

The enthalpy of ion formation in aqueous solution relative to the new reference level is determined by the Helmholtz equation (3.5):

$$\left(\Delta H_{\text{ion}}^{0}\right)^{*} = \left(\Delta G_{\text{ion}}^{0}\right)^{*} - T \left(\frac{\partial \varphi^{*}}{\partial T}\right)_{p} \tag{3.10}$$

The relationships between the tabulated values of enthalpy and free enthalpy and our values in the new reference frame are

$$\left(\Delta H^0_{\text{ion}}\right)^* = \Delta H^0_{\text{ion}} \pm \frac{n_e}{2} \Delta H^0_{\text{H}_2\text{O}} \tag{3.11}$$

$$\left(\Delta G^0_{\text{ion}}\right)^* = \Delta G^0_{\text{ion}} \pm \frac{n_e}{2} \Delta G^0_{\text{H}_2\text{O}} \,. \tag{3.12}$$

In (3.9, 11, 12) the plus sign indicates cations, the minus sign is for anions.

From these quantities we can now find the chemical energy and exergy. Since the chemical energy and exergy of ions in infinitely dilute aqueous solutions are set to zero, by analogy with (3.3, 4) we can write

$$\Sigma_j m_j i^0_j = -\left(\Delta H^0_{\text{ion}}\right)^* \tag{3.13}$$

$$\Sigma_j m_j e^0_j = -\left(\Delta G^0_{\text{ion}}\right)^* \,. \tag{3.14}$$

When the ion of a pure element is considered, the chemical energy and exergy are equal to the enthalpy and Gibbs energy of ion formation in aqueous solutions, i.e.,

$$i^0_j = -\left(\Delta H^0_{\text{ion}}\right)^* \tag{3.15}$$

$$e^0_j = -\left(\Delta G^0_{\text{ion}}\right)^* \,. \tag{3.16}$$

When the ion is composed of several elements, the values of chemical energy and exergy of one of the elements can be found by (3.13, 14), if these quantities are known for the other elements.

We shall now illustrate the procedure by a few examples. For aluminum, we take the values of the enthalpy and the Gibbs energy of forming of $Al^{3+}$ in aqueous solution from [43], and apply (3.11, 12) and (3.15, 16). From (3.11, 12) we find

$$\Delta H^* = +\left[-524.7 + \tfrac{3}{2}(-286)\right] = -953.7 \text{ kJ/g-ion}$$

$$\Delta G^* = +\left[-481.2 + \tfrac{3}{2}(-237)\right] = -836.7 \text{ kJ/g-ion} \,.$$

Then according to (3.15, 16) the chemical energy and exergy of aluminum are

$$i_{\text{Al}} = 953.7 \text{ kJ/mole} \quad \text{and} \quad e_{\text{Al}} = 836.7 \text{ kJ/mole} \,.$$

The value of chemical exergy of aluminum can also be determined from the standard electrode potential. Taking the value for the standard potential of formation of aluminum ion in its 3+ oxidation state from [43], according to (3.8) we have

$$\varphi^*_{\text{Al}^{3+}} = [-1.663 + (-1.23)] = -2.893 \text{ V} \,.$$

Then by (3.9) we determine

$$\Delta G^* = -2.893 \cdot 3 \cdot 96.5 = -837 \text{ kJ/g-ion} \,.$$

The values of chemical energy and exergy of negatively charged ions are calculated in a similar manner. For example, using the values of enthalpy and free enthalpy of forming the ion of chlorine ($Cl^-$) [43], we have

$$\Delta H^* = \left[-167.46 - \tfrac{1}{2}(-286)\right] = -24.46\ \text{kJ/g-ion}\ ,$$
$$\Delta G^* = \left[-131.17 - \tfrac{1}{2}(-237)\right] = -12.67\ \text{kJ/g-ion}\ .$$

Hence, from (3.15, 16) we obtain

$$i_{Cl_2} = 48.92\ \text{kJ/mole} \quad \text{and} \quad e_{Cl_2} = 25.34\ \text{kJ/mole}\ .$$

Finally, we shall show how the chemical energy and exergy of elements (or elementary substances) are calculated using (3.3, 4). For example, with the values of $\Delta H_{aq}$ and $\Delta G_{aq}$ for orthophosphoric acid [65] we find the chemical energy and exergy of phosphorus. For this we need to know the chemical energy and exergy of the other elements in orthophosphoric acid, oxygen and hydrogen. For hydrogen $i_{H_2}$ = 286 kJ/mole, $e_{H_2}$ = 237 kJ/mole and for oxygen these values are equal to zero. Then by (3.3, 4)

$$i_P = \left(\Delta H_{H_3PO_4}\right)_{aq} - \Sigma m_j i_j = 1295 - \tfrac{3}{2}286 = 866\ \text{kJ/mole}\ ,$$
$$e_P = \left(\Delta G_{H_3PO_4}\right)_{aq} - \Sigma m_j e_j = 1150 - \tfrac{3}{2}237 = 795\ \text{kJ/mole}\ .$$

Thus, by taking infinitely dilute aqeous solutions of ions in a high oxidation state as a reference level we can make use of [32, 43, 65, 85, 100, 101, 126, 128, 133, 171, 189] that are not applicable in the techniques of other authors. Among them are the change in the enthalpy and free enthalpy of forming the infinitely dilute water solutions, the enthalpy and free enthalpy of forming substances in aqueous solutions, the enthalpy and free enthalpy of forming ions in water solutions and, finally, the standard electrode potentials. These quantities are interrelated by

$$\Delta H_z^0 + (\Delta H_{dis})_z = m_1\left(\Delta H_c^0\right)_z + m_2\left(\Delta H_a^0\right)_z = \left(\Delta H_z^0\right)_{aq} \tag{3.17}$$
$$\Delta G_z^0 + (\Delta G_{dis})_z = m_1\left(\Delta G_c^0\right)_z + m_2\left(\Delta G_a^0\right)_z = \left(\Delta G_z^0\right)_{aq} \tag{3.18}$$

where $(\Delta G_{dis})_z$ is the change in the Gibbs energy (free enthalpy) in forming the infinitely dilute aqueous solution of a substance $z$. These relations allow one to ensure that the values of the chemical energy and exergy of substances calculated from different thermodynamic quantities are self-consistent.

This author has used this method to calculate the chemical energy and exergy of elements and elementary substances using standard thermodynamic data. The values obtained using (3.17, 18) are reported in Table 3.1.

Using the data of Table 3.1 supplemented by the thermodynamic constants given in [32, 43, 65, 85, 100, 101, 126, 128, 133, 171, 189] the chemical energy and exergy for a large number of inorganic and organic compounds have been calculated by using (2.14, 15) and published in [153]. These results allow the application of the exergy method to further industrial production processes.

**Table 3.1.** Comparison of specific chemical energy and exergy obtained by different techniques [kJ/mole]

| Number of element in periodic systems | Symbol | from data: [182, 183, 186] Reference species | Chemical exergy | [201] Reference species [for exergy] | Theta function | Reference exergy | [153] Reference species [for exergy] | Chemical energy | Chemical exergy |
|---|---|---|---|---|---|---|---|---|---|
| 1. | H | $H_2O$ (gas) | 119.18 | $H_2O$ (liq) | 143.0 | 117.7 | $H_2O$ (liq) | 143 | 118.5 |
| 2. | He | He | 30.36 | He | 0.0 | 30 | He | 0.0 | 0.0 |
| 3. | Li | $Li^+$ | 393.0 | LiCl<br>[$LiCl \cdot H_2O$] | 408.5 | 372.2 | $Li^+$ | 455 | 427 |
| 4. | Be | | | $BeSO_4 \cdot 4H_2O$<br>[$BeO \cdot Al_2O_3$] | 643.3 | 594.6 | $Be^{2+}$ | 675 | 594 |
| 5. | B | $B(OH)_3$ (aq)[1)] | 628.5 | $H_3BO_3$ | 666.0 | 610.7 | $B_2O_3$ | 715 | 675 |
| 6. | C | $HCO_3^-$ | 408.5 | $CO_2$ | 393.8 | 410.8 | $CO_2$ | 394 | 395 |
| 7. | N | $N_2$ | 0.36 | air | 0.0 | 0.3 | $N_2$ | 0.0 | 0.0 |
| 8. | O | $O_2$ | 2.0 | air | 0.0 | 2.0 | $O_2$ | 0.0 | 0.0 |
| 9. | F | $F^-$ | 233.15 | $Ca_{10}(PO_4)_6 \cdot F_2$ | 254.4 | 308.4 | $F^-$ | 186 | 158 |
| 10. | Ne | Ne | 27.15 | Ne | 0.0 | 27.1 | Ne | 0.0 | 0.0 |
| 11. | Na | $Na^+$ | 336.6 | $NaNO_3$ | 468.2 | 361.0 | $Na^+$ | 423 | 381 |
| 12. | Mg | $CaCO_3 \cdot MgCO_3$ | 633.8 | $CaCO_3 \cdot MgCO_3$ | 726.4 | 618.6 | $Mg^{2+}$ | 762 | 693 |
| 13. | Al | $Al_2SiO_5$ | 887.89 | $Al(NO_3)_3 \cdot 6H_2O$<br>[$Al_2O_3$] | 1136.2 | 788.8 | $Al^{3+}$ | 954 | 837 |
| 14. | Si | $SiO_2$ | 803.01 | $SiO_2$ | 911.6 | 853.3 | $SiO_2$ | 910 | 860 |
| 15. | P | $Ca_3(PO_4)_2$ | 875.8 | $Ca_3(PO_4)_2$ | 840.9 | 866.5 | $H_3PO_4$ (aq) | 866 | 795 |
| 16. | S | $SO_4^{2-}$ | 609.6 | $CaSO_4 \cdot 2H_2O$ | 638.0 | 603.2 | $H_2SO_4$ (aq) | 624 | 523 |
| 17. | Cl | $Cl^-$ | 61.8 | $Cl_2$<br>[NaCl] | 0.0 | 23.5 | $Cl^-$ | 25 | 13 |
| 18. | Ar | Ar | 11.67 | Ar | 0.0 | 11.7 | Ar | 0.0 | 0.0 |
| 19. | K | $K^+$ | 366.6 | $KNO_3$ | 495.0 | 387.1 | $K^+$ | 452 | 410 |
| 20. | Ca | $Ca^{2+}$ | 712.4 | $CaCO_3$ | 814.0 | 712.8 | $Ca^{2+}$ | 853 | 797 |

(continued)

| | | | | | | | | | |
|---|---|---|---|---|---|---|---|---|---|
| 21. | Sc | | | $Sc_2O_3$ | 955.0 | 907.4 | $Sc^{3+}$ | 1055 | 958 |
| 22. | Ti | $TiO_2$ | 876.0 | $TiO_2$ | 945.4 | 886.2 | $TiO_2$ | 945 | 890 |
| 23. | V | $V_2O_5$ | 725.88 | $V_2O_5$ | 775.8 | 705.4 | $V_2O_5$ | 785 | 715 |
| | | | | $Cr_2O_3$ | | | | | |
| 24. | Cr | $Cr_2O_3$ | 538.61 | $[K_2Cr_2O_7]$ | 570.2 | 547.8 | $Cr^{3+}$ | 658 | 572 |
| 25. | Mn | $MnO_2$ | 482.3 | $MnO_2$ | 520.4 | 461.6 | $Mn^{2+}$ | 522 | 465 |
| 26. | Fe | $Fe_2O_3$ | 377.74 | $Fe(NO_3)_3 \cdot 9H_2O$ | 713.3 | 368.4 | $Fe^{3+}$ | 477 | 370 |
| | | | | $[Fe_2O_3]$ | | | | | |
| 27. | Co | $Co_3O_4$ | 265.0 | $CoCl_2 \cdot 6H_2O$ | 400.7 | 288.6 | $Co^{2+}$ | 354 | 292 |
| | | | | $[CoFe_2O_4]$ | | | | | |
| 28. | Ni | $Ni^{2+}$ | 232.7 | $NiCl_2 \cdot 6H_2O$ | 388.4 | 243.6 | $NiSO_4 \cdot 7H_2O$ | 365 | 326 |
| 29. | Cu | $Cu^{2+}$ | 134.2 | $CuCl_2 \cdot 2H_2O$ | 249.8 | 143.9 | $Cu^{2+}$ | 225 | 175 |
| | | | | $[Cu_4(OH)_6Cl_2]$ | | | | | |
| 30. | Zn | $Zn^{2+}$ | 339.2 | $Zn(NO_3)_2 \cdot 6H_2O$ | 592.0 | 337.7 | $Zn^{2+}$ | 458 | 388 |
| 31. | Ga | | | $Ga_2O_3$ | 544.9 | 496.5 | $Ga^{3+}$ | 580 | 510 |
| 32. | Ge | | | $GeO_3$ | 551.4 | 493.4 | $H_2GeO_3$ (aq) | 615 | 570 |
| 33. | As | $HAsO_4^{2-}$ | 494.6 | $As_2O_3$ | 462.8 | 386.4 | $H_3AsO_4$ (aq) | 523 | 392 |
| 34. | Se | $SeO_4^{2-}$ | 346.5 | $H_2SeO_4 \cdot H_2O$ | 269.1 | 167.4 | $Se^{2+}$ | 440 | 415 |
| | | | | $[SeO_2]$ | | | | | |
| 35. | Br | $Br^-$ | 50.6 | $PtBr_2$ | 24.3 | 34.4 | BrO (aq) | 98 | 34 |
| 36. | Kr | Kr | 34.32 | | | | Kr | 0.0 | 0.0 |
| 37. | Rb | $Rb^+$ | 388.6 | $RbNO_3$ | 495.4 | 389.8 | $Rb^+$ | 423 | 407 |
| 38. | Sr | $Sr^{2+}$ | 730.2 | $Sr(NO_3)_2 \cdot 4H_2O$ | 1012.1 | 771.7 | $SrSO_4$ | 875 | 811 |
| | | | | $[SrCl_2 \cdot 2H_2O]$ | | | | | |
| 39. | Y | | | $YF_3$ | 956.6 | 933.1 | $Y^{3+}$ | 1130 | 1040 |
| | | | | $[Y(OH)_3]$ | | | | | |
| 40. | Zr | | | $ZrSiO_4$ | 1122.9 | 1059.3 | $Zr^{4+}$ | 1170 | 1065 |
| 41. | Nb | | | $Nb_2O_5$ | 950.4 | 878.7 | $Nb_2O_5$ | 950 | 882 |
| 42. | Mo | $MoO_4^{2-}$ | 730.3 | $CaMoO_4$ | 728.5 | 714.9 | $H_2MoO_4$ (aq) | 797 | 728 |
| 43. | Tc | | | | | | $TcO_4^-$ | 580 | 512 |
| 44. | Ru | | | Ru | 0.0 | 0.0 | $RuO_2$ | 297 | 251 |

(continued)

| | | | | | | | | | |
|---|---|---|---|---|---|---|---|---|---|
| 45. | Rh | | | Rh | 0.0 | 0.0 | $Rh^{3+}$ | 151 | 126 |
| 46. | Pd | | | Pd | 0.0 | 0.0 | $Pd(ReO_4)_2$ | 226 | 126 |
| | | | | | | | $[PdCl_2]$ | | |
| 47. | Ag | $AgCl_2^-$ | 70.2 | AgCl | 127.2 | 86.4 | $Ag_2MoO_4$ | 85 | 50 |
| 48. | Cd | $CdCl_2$ (aq) | 293.8 | $Cd(NO_3)_2$ | 456.6 | 304.4 | $Cd^{2+}$ | 358 | 315 |
| | | | | $[CdCl_2 \cdot 5/2H_2O]$ | | | | | |
| 49. | In | | | $In_2O_3$ | 463.2 | 412.7 | $In^{3+}$ | 610 | 455 |
| 50. | Sn | $SnO_2$ | 542.66 | $SnO_2$ | 581.1 | 516.1 | $SnO_2$ | 594 | 518 |
| 51. | Sb | $Sb_2O_5$ | 359.19 | $Sb_2O_5$ | 486.3 | 410.0 | $Sb_2O_5$ | 504 | 432 |
| 52. | Te | | | $TeO_2$ | 322.8 | 266.5 | $KTeO_4$ | 568 | 510 |
| 53. | I | $IO_3^-$ | 87.35 | $KIO_3$ | 6.7 | 25.6 | $IO_3^-$ | 87 | 17 |
| 54. | Xe | Xe | 40.3 | | | | Xe | 0.0 | 0.0 |
| 55. | Cs | $Cs^+$ | 404.4 | CsCl | 443.1 | 391.2 | $Cs^+$ | 518 | 415 |
| 56. | Ba | $Ba^{2+}$ | 747.7 | $Ba(NO_3)_2$ | 992.7 | 784.7 | $BaSO_4$ | 858 | 830 |
| 57. | La | | | $LaCl_3 \cdot 7H_2O$ | 1178.6 | 983.2 | $La^{3+}$ | 1135 | 1090 |
| 58. | Ce | | | $CeO_2$ | 1089.4 | 1021.4 | $Ce^{3+}$ | 1130 | 1075 |
| 59. | Pr | | | $PrCl_3 \cdot 7H_2O$ | 1177.7 | 926.8 | $Pr^{3+}$ | 1130 | 1070 |
| | | | | $[Pr(OH)_3]$ | | | | | |
| 60. | Nd | | | $NdCl_3 \cdot 6H_2O$ | 1160.2 | 967.7 | $Nd^{3+}$ | 1115 | 1060 |
| 61. | Pm | | | | | | $Pm^{3+}$ | 1135 | 1060 |
| 62. | Sm | | | $SmCl_3 \cdot 6H_2O$ | 1156.0 | 963.5 | $Sm^{3+}$ | 1335 | 1260 |
| 63. | Eu | | | $EuCl_3 \cdot 6H_2O$ | 1070.6 | 873.1 | $Eu^{3+}$ | 1160 | 1050 |
| 64. | Gd | | | $GdCl_3 \cdot 6H_2O$ | 1151.8 | 958.9 | $Gd^{3+}$ | 1110 | 1050 |
| 65. | Tb | | | $TbCl_3 \cdot 6H_2O$ | 1145.1 | 948.0 | $Tb^{3+}$ | 1125 | 1050 |
| 66. | Dy | | | $DyCl_3 \cdot 6H_2O$ | 1156.0 | 958.9 | $Dy^{3+}$ | 1115 | 1040 |
| 67. | Ho | | | $HoCl_3 \cdot 6H_2O$ | 1164.0 | 967.3 | $Ho^{3+}$ | 1105 | 1030 |
| 68. | Er | | | $ErCl_3 \cdot 6H_2O$ | 1160.2 | 961.4 | $Er^{3+}$ | 1090 | 1020 |
| 69. | Tm | | | $Tm_2O_3$ | 945.0 | 894.9 | $Tm^{3+}$ | 1090 | 1015 |
| 70. | Yb | | | $YbCl_3 \cdot 6H_2O$ | 1131.7 | 936.3 | $Yb^{3+}$ | 1130 | 1010 |
| 71. | Lu | | | $LuCl_3 \cdot 6H_2O$ | 1116.7 | 918.3 | $Lu^{3+}$ | 1060 | 1010 |
| 72. | Hf | | | $HfO_2$ | 1145.5 | 1023.9 | $Hf^{4+}$ | 1185 | 1130 |

(continued)

| | | | | | | | | | |
|---|---|---|---|---|---|---|---|---|---|
| 73. | Ta | | | $Ta_2O_5$ | 1023.7 | 951.3 | $Ta_2O_5$ | 1045 | 974 |
| 74. | W | $WO_4^{2-}$ | 827.5 | $CaWO_4$ | 832.3 | 818.8 | $H_2WO_4$ | 855 | 800 |
| 75. | Re | | | | | | $ReO_4^-$ | 648 | 585 |
| 76. | Os | | | $OsO_4$ | 394.4 | 297.4 | $OsO_4$ | 394 | 303 |
| 77. | Ir | | | Ir | 0.0 | 0.0 | $IrO_2$ | 280 | 188 |
| 78. | Pt | | | Pt | 0.0 | 0.0 | $PtO_2$ | 134 | 84 |
| 79. | Au | $AuCl_2^-$ | 15.4 | Au | 0.0 | 0.0 | $Au(OH)_3$ | 49 | 15 |
| | | | | | | | [$AuCl_3$] | | |
| 80. | Hg | $HgCl_4^{2-}$ | 115.9 | $HgCl_2$ | 224.4 | 131.8 | $Hg^{2+}$ | 112 | 73 |
| 81. | Tl | | | $TlCl_3$ | 315.3 | 169.8 | $Tl_2O_4$ | 239 | 181 |
| | | | | [$Tl_2O_4$] | | | | | |
| 82. | Pb | $PbCl_2$ | 232.8 | $PbCl_2$ | 359.6 | 337.5 | $PbWO_4$ | 320 | 305 |
| | | | | [PbCl(OH)] | | | | | |
| 83. | Bi | $BiO^+$ | 274.5 | $BiCl_3$ | 379.3 | 296.9 | $Bi^{3+}$ | 370 | 294 |
| | | | | [BiOCl] | | | | | |
| 86. | Rn | | | | | | Rn | 0.0 | 0.0 |
| 88. | Ra | | | | | | $RaCO_3$ | 880 | 840 |
| | | | | | | | [$RaSO_4$] | | |
| 89. | Ac | | | | | | $Ac_2O_3$ | 1025 | 980 |
| 90. | Th | | | $ThO_2$ | 1227.2 | 1165.6 | $Th^{4+}$ | 1345 | 1260 |
| 92. | U | $UO_3$ | 1224.2 | $U_3O_8$ | 1192.4 | 1122.9 | $U_3O_8$ | 1260 | 1180 |
| 93. | Np | | | | | | $Np^{4+}$ | 1160 | 995 |
| 94. | Pu | | | | | | $PuO_2$ | 1150 | 1025 |
| 95. | Am | | | | | | $Am^{3+}$ | 1111 | 1026 |

Note: [1] Aqueous solution

### 3.2.2 Comparison of the Different Methods

Table 3.1 also lists the values of chemical energy and exergy obtained by other means. Although the quantities given, the enthalpy of devaluation $d_n$, the $\Theta$-function and the chemical energy of an element $i_{ch}$ are identical by definition, the determination of the reference exergy and the assumptions made do not coincide. In the normal chemical exergy of all elements takes into account the concentration component (residual exergy, discussed below). Yamauchi and Fueki include this component in the reference exergy only for those elements whose reference species are atmospheric components while the authors of [120, 153] do not consider residual exergy at all. Therefore, in addition to the values of the chemical exergy according to Szargut, Table 3.1 also presents the values of the free enthalpy of devaluation, which corresponds to our chemical exergy $e_{ch}$.

Table 3.1 shows that in most cases the chemical energy and exergy determined by the Ozoling–Stepanov technique is larger than those found by other methods. This confirms that the authors of this technique managed to select the lowest energy reference level of the considered characteristics.

In some cases the reference species are the same but the chemical energy (exergy) determined by different techniques are different. This is due to the differences (sometimes rather large) in the values of thermodynamic reference data accepted in different publications. This demonstrates that the use of several different thermodynamic quantities for calculating the chemical energy and exergy which allows a check on the self-consistency of the final results [120, 153], is undoubtedly better than the alternative methods.

We now return to the issue of inclusion of the "residual exergy". This quantity can be used to take into account the exergy releases from an industrial plant into the environment as well as to evaluate the minimum amount of work required to extract material and energy resources from the environment.

Although we share the aspiration of some authors [115, 173, 182–184, 186] to account for the differing distributions of elements in the environment when determining the chemical exergy, it is difficult to agree with their suggested approaches. Rather complex expressions and not always convincing initial assumptions have been suggested for the determination of the residual exergy value. We believe that because reference energy values are used in the analysis of industrial processes, in determining these limiting quantities it is necessary to approximate real conditions as closely as possible. For example, no real process which involves the transformation of a natural mineral or energy source obtains them from the environment in the same way that extraction of a component from a solution occurs. Even real processes in the atmospheric separation are based on completely other principles. Another example is the production of the mineral raw material from sea water via evaporation by solar radiation that is not taken into account in the selection of a reference system.

This fact is more obvious in the processes of extraction of mineral and energy resources from the earth crust. These usualy assume that the lithosphere is not an ideal solution and that there are areas of local higher concentration components (deposits) than their average content in the used environment models. But energy and material resources are extracted precisely from such deposits. The amount of work (energy) consumed per unit of product output is chiefly determined in this case by the depth and conditions of occurence useful minerals of these deposits and their surroundings in the earth's crust rather than by the content of the deposit itself.

Thus we conclude that calculating the residual exergy is not that important. One of the main reasons is the fact that its value is usually within the error limits the thermodynamic data used to calculate the chemical energy and exergy. Table 3.1 shows this rather convincingly.

At the same time one should not neglect the fact that different interpretations of the concept of chemical exergy of substances undoubtedly hampers its wider use in practical applications. Therefore, it seems reasonable to develop a common viewpoint on this issue. For example, while not denying the necessity of including residual exergy of substance in the analysis of some processes (such as separation of atmospheric components, extraction of products from aqueous solutions, etc.), one should be able to agree that, in general, it is unnecessary to include it in the values of chemical exergy of elements. The use of quantities taking this component into account in industrial processes with complex chemical transformations leads to results that are not comparable with those obtained by the methods of classical thermochemistry. This does not argue for extensive introduction of these otherwise very useful thermodynamic quantities into daily practice.

# 4. Optimizing the Use of Thermal Secondary Energy Resources

One of the most important contributors to fuel conservation in industry is the utilization of secondary energy resources that inevitably appear in production processes [5, 15, 24, 42, 73, 118, 124, 141, 188, 192, 194, 196]. Secondary energy resources imply an energy potential of products and wastes formed in technological processes (aggregates) that is not used fully in the primary considered process, but can be used as a source of energy supply of other consumers. Depending on the form of this potential the secondary energy resources are divided into three principal groups: fuel, thermal, and hyper-pressure sources.

The use of secondary energy resources of the first and third groups presents no difficulties nowadays, since the technological and economic problems of their utilization have in general been solved. Hence, in the discussion below, detailed consideration will be given to the use thermal secondary energy resources, including the heat of products, intermediate materials, and wastes that cannot be used in the primary process.

## 4.1 Thermal Secondary Energy Resources

Strictly speaking, we should simultaneously solve the problem of determining the most advantageous extent of using secondary energy resources and the problem of choosing the areas where they should be utilized. Based on the availability of heat consumers in the area and other factors, the thermal secondary energy resources can be used in one of the following regimes:

a) thermal, when steam and hot water produced by these resources are used as the heat supply;
b) electrical,
c) combined, with production of thermal power and electricity in the heat recovery installations (heat recovery cogeneration plants).

In principle, the problem of determining the most advantageous extent of using secondary energy resources should be solved for each of the given regimes. Then the one resulting in the largest economic gain can be chosen. However, it is sometimes the preliminary analysis that reveals the most expedient regime of utilizing secondary energy resources at a plant.

In each specific case this problem is approached by considering the effect of several undesirable factors: relatively low values of available capacity of each existing source of secondary energy; comparatively low potential of secondary energy produced; unfavorable conditions for utilization due to high dust content and corrosive capability of flue gases from production units, etc. These aspects of secondary energy sources arise because the presently available heat recovery installations are lower in efficiency, amounts of generated steam (hot water) and hence other indices than common power installations using primary energy carriers. Therefore, their use is considerably less efficient in general, and particularly for electricity production. For the most efficient devices available the ratio of the efficiencies of a waste heat boiler and a boiler burning primary fuel is about 0.75 – 0.80, and the ratio of the efficiencies of small heat recovery condensation power plants working at low steam parameters and large condensation power plants is 0.4 – 0.6. Thus, secondary energy sources are less efficient at electricity production, that is, they require a greater net amount of fuel. Moreover, the capital investment required to construct small power plants and the operating costs are more than twice those for large condensation power plants. In fact, the use of secondary energy resources for electricity production at small heat recovery power plants is usually the *least* economically efficient route. Experiments on power production in the USSR have shown that thermal secondary energy resources are best used for the thermal production of hot water and steam. This is the application which is considered in this book in greater detail.

Nevertheless, when the demands for heat are limited, it may indeed appear expedient to utilize secondary energy resources in the combined route and in case demands are completely absent they could be used in electricity production. The increase in utilization costs in these cases can be offset by the more complete consumption of the available secondary energy resources. Examples of the combined scheme exist in many countries.

The contribution of heat recovery installations to meeting an industrial plant's own load curve as well as that of the adjoining region leads to the conservation of fuel for heat production in the energy supply system (ESS) of the region.[1] Note, however, that the rate of conservation will go down as the capacity of commissioned waste-heat installations increases. This is because their capacity will be considerably underused due to the diversity of the heat load curves of the region, increasing costs on heat networks as well as other factors. Thus, on a national level this will lead to higher material and financial expenditures, since the specific capital investment required for waste-heat boilers is much higher than for power boilers using primary fuel.

Thus, the decreasing economic effect of secondary energy resource utilization and its increasing costs suggest that there should be an optimal alternative corresponding to the maximum economic benefit. This is illustrated by the graph in

[1] The authors assume that the proposed model and algorithms for optimizing the extent of secondary energy resources utilization of a single industrial plant are completely suitable for solving similar problems in an industrial complex of plants.

Fig. 4.1. The curve EF represents the economic effect in the energy supply system as a function of the secondary energy resources of the capacity $L_{\mathrm{SER}}$ in meeting the heat load curve in a given region. The curve $\Delta Z$ is the growth of expenditures in secondary energy resources utilization as a function of the total capacity of heat recovery installations. The point $EF_{\mathrm{res}}$ = max determines the most advantageous value from the economic standpoint of utilized secondary energy resources. Further growth in capacity $L_{\mathrm{SER}} > L_{\mathrm{SER}}^{\mathrm{opt}}$ will lead to decrease in the economic benefit and at some value $L'_{\mathrm{SER}}$, $EF_{\mathrm{res}}$ will be equal to zero. At such thermal outputs at recovery installations, alternative energy supplies will be of equal economic efficiency. Utilization of secondary energy resurces above the value $L'_{\mathrm{SER}}$ will result in expenditure overrun.

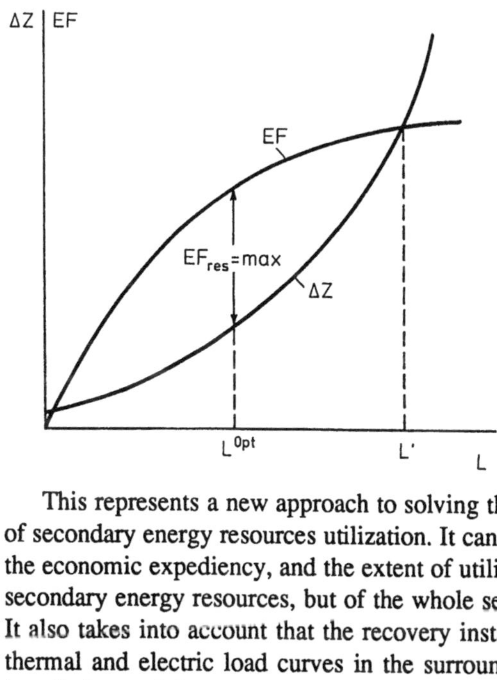


**Fig. 4.1.** Economic effect in the energy system as a function of capacity of heat recovery installations engaged in meeting the thermal load curve of a region

This represents a new approach to solving the problem of economic efficiency of secondary energy resources utilization. It can be applied to the determination of the economic expediency, and the extent of utilization of not individual sources of secondary energy resources, but of the whole set of sources available at the plant. It also takes into account that the recovery installations contribute to meeting the thermal and electric load curves in the surrounding region in parallel with other installations of the energy supply system. This enables the determination of the optimal composition of generating capacities of the energy supply system and hence, the optimal extent of secondary energy resources utilization at the plant.

This general problem requires a series of smaller problems to be solved. One of them is to derive a procedure for determining the minimal costs on secondary energy resources utilization that correspond to the optimal composition of recovery installations for any possible volume of power supplied by the plant to the energy supply system of the region.

## 4.2 Minimizing Costs. Optimal Composition of Heat Recovery Installations

The costs of utilizing secondary energy resources determine the choice of the optimal extent of their use at a plant. Difficulties in the cost estimation are first, the necessity to calculate the costs as a function of the available capacity (output) of recovery installations which is dependent on operating conditions and the reliability of production units. Second, it is necessary to solve an optimization problem, since at any plant there are usually a great number of sources of secondary energy resources which are quite varied in terms of the conditions necessary for utilization and costs. Hence, for any current value of the capacity $L_{SER}$, which contributes to meeting the thermal and electric load curves of the region, the best composition of recovery installations, each contributing its optimum amount of secondary heat, must be chosen so that the utilization costs for the whole plant are minimal. In other words, this requires the determination of an equivalent economic curve (relation for the entire set of available secondary energy resources $\underset{\min}{Z}_{SER} = f(L_{SER})$, each point of which corresponds to the optimal "internal" parameters of the most advantageous composition and output of individual recovery installations.

Before we continue with detailed consideration of the statement of the problem and methods of solution let us emphasize the initial premises. To determine the costs of utilization for each source, it is necessary to proceed from the idea that the construction of the recovery installation should not result in decrease of output, violation of production schedules or deterioration of working conditions in both the primary production process and other aggregate connected to it. It is assumed that the net costs of utilization of secondary energy resources of any source are determined as the difference between the costs of utilization of secondary energy and of no utilization, i.e.

$$\Delta Z_i = Z_{1_i} - Z_{2_i} \tag{4.1}$$

where $Z_{1_i}$ are the costs of source $i$ with secondary energy recovery and $Z_{2_i}$ is the same without recovery. The determination of the costs of each source $i$ of secondary energy resources, considering its operating conditions, reliability of recovery installations, and other factors will be considered in the next section.

### 4.2.1 Costs of Production of Secondary Energy Resources

The amount of heat produced by individual recovery installations can be highly diverse, since they are determined by the operating conditions of the primary aggregates. The total heat production curve by the plant will be non-uniform (irregular) as well. It should be taken into account that in optimization calcuations the composition of heat recovery installations changes with every change in the total value of the secondary energy resources used. Therefore, the curve of the total secondary energy production is characterized by a quite definite non-uniformity (irregularity) which depends on the chosen composition of the recovery installations.

Thus, the determination of the optimal degree of secondary energy resource utilization turns out to be extremely complicated. Moreover, practical implementation of the solution would lead to highly complex operating conditions of cogeneration plants, regional and industrial boilers. In practice, excessive sophistication in energy producing equipment working in tandem with recovery installations can be avoided by levelling off the heat production curve by using supplementary standard furnace facilities.

One should also note that the largest effect of secondary energy resources use (the largest amount fuel conserved) can be achieved by allowing the waste heat of the plant to supply most of the total heat load curve for the considered region.

Thus, levelling off the heat production curve of the plant can be solved in two ways. The first is to level off the curve for each recovery installation by construction of special furnace facilities. The second way implies levelling off the total heat production curve via construction of special units at the cogeneration plant or boiler. Usually the first method was practiced, i.e., recovery installations designed for parallel work with production aggregates of batch operation were equipped with special furnace or storage facilities. Therefore, this method will be discussed below.

For the variable curve of the energy output by the aggregate, the optimal capacity of the recovery installation must be calculated with regard for the costs of levelling the curve, i.e., costs of the construction of the supplementary furnace or storage facilities and costs of their operation (cost of the fuel burnt, maintenance, etc.). In addition, account should be taken of the costs of the additional cooling installations that are needed if the capacity of the recovery installation is chosen to be lower than the maximum hourly output of secondary resources by the technological aggregate. In accordance with the requirements of the employed technology, waste gases are cleaned from dust and harmful impurities or/and valuable components are extracted from them. In such a case the waste gases must be cooled to some limiting temperature appropriate for the gas cleaning procedure. This temperature must be provided whether or not the heat contained in the secondary energy resources is utilized usefully by the recovery installations or is removed by other cooling mechanisms into the environment. If these conditions are neglected, a large error can appear in the calculated optimal volume of secondary energy resources used.

The expression for the costs of a set of recovery, cooling and furnace installations depends on the method of variation of the heat output of recovery installations. In general the method for output variation is determined by the arrangement of the recovery and cooling systems. Two arrangements are possible: In the first, the installations are placed in series. In this case the total volume of gas at a known temperature passes through both the recovery and cooling systems. The capacity of the recovery installation can be changed by varying the temperature at the recovery installation outlet and at the cooling installation inlet.

In the second arrangement, the recovery and cooling installations are arranged in parallel, i.e., the total gas flow is separated into two flows: one of them passes through the recovery installations, the other through the cooling system. The flows are cooled to some limiting temperature $t_{\text{lim}}$ which is determined by the method of gas purification.

However, to ensure reliable operation of the production aggregates, preference is given to the second, parallel arrangement. In this case the heat output can be varied primarily by changing the relative volumes of gases passing through the recovery and cooling blocks. Two methods are possible here: changing the cross section of the gas duct in the recovery installation and thus the size of the whole installation at a constant gas velocity or changing the gas velocity by regulating the vacuum along the waste gas path.

In the first case, volume variation, the heat production by the recovery installations will be a function of the relative gas volume and hence, also of the secondary energy contained in the gas that will pass through the recovery and cooling systems. In the second case, velocity variation, the curve of heat produced by the recovery installation will approximately correspond to the curve of secondary energy resources output by the aggregate and only the maximum capacity of the recovery installation will change. Thus, the expression for the costs of secondary resources utilization in a source $i$ can be written

$$Z_1 = Z_\text{r} + Z_\text{c} + Z_\text{f} \tag{4.2}$$

where $Z_\text{r}$ are the costs of construction and operation of recovery installations; $Z_\text{c}$ is the same for cooling systems and $Z_\text{f}$ is that for furnace facilities. $Z_\text{c}$ and $Z_\text{f}$ are functions of the form of curves for secondary energy output by each source.

The costs of heat recovery installations comprise the costs of their construction and operation. Their dependence on the installation capacity can be written

$$Z_\text{r} = \varepsilon_\text{r} K_\text{r}(L_\text{r}) \tag{4.3}$$

where $\varepsilon_\text{r} = \varepsilon_{\text{dep}} + \varepsilon_{\text{rr}} + \varepsilon_{\text{ser}} + \varepsilon_\text{s}(\varepsilon_{\text{dep}}, \varepsilon_{\text{rr}}, \varepsilon_{\text{ser}}$ are the costs of depreciation, routine repair and servicing of the recovery installations, respectively; $\varepsilon_\text{s}$ is an interest on capital) and $K_\text{r}(L_\text{r})$ are the capital investment in the recovery installations as a function of their capacity.

The costs of cooling systems are also a function of their capacity which can be found assuming unchanged operating conditions of the gas cleaning devices:

$$L_\text{c}^* = L^* - L_\text{r} \tag{4.4}$$

where $L^*$ is the maximum volume of output of secondary energy resources by the technological aggregate and $L_\text{c}^*$ is the capacity of the cooling system. Then,

$$Z_\text{c} = \varepsilon_\text{c} K_\text{c}(L_\text{c}) \tag{4.5}$$

where $\varepsilon_\text{c}$ are the total costs of depreciation, routine repair and servicing of the cooling systems in fractions of the capital investment including the interest on capital $\varepsilon_s$; $K_c(L_c)$ is the capital investment in the cooling system as a function of its capacity.

It has already been noted that the curve of heat output can be levelled off by using special supplementary furnace facilities present in recovery installations, if necessary. The costs of facilities are again a function of the costs of their construction and operation and the cost of fuel burnt:

$$Z_f = \varepsilon_f K_f(L_f^*) + a_f \beta_f Q_f \tag{4.6}$$

where $\varepsilon_f$ is again the total cost of depreciation, routine repair and servicing of the furnace facilities in fractions of capital investment including the interest on capital $\varepsilon_s$, $K_f(L_f^*)$ is the capital investment in the furnace facilities as a function of their capacity $L_f^*$; $Q_f$ is the total output of the furnace facilities; $a_f$, $\beta_f$ are the cost of fuel burnt in the furnace installations and its specific consumption per Gcal (GJ) of heat produced, respectively.

The first cost component is a function of the capacity of furnace facilities which in turn depends on the curve of secondary energy resources output and the capacity of the recovery installations. The second component is a function of the specific consumption and the cost of fuel and heat production by the furnace facilities in the considered time interval. In this case the latter depends on the curve of secondary energy resources output and on the variable value of capacity of the recovery installation.

If the curve of secondary energy resources output of source $i$ is given by a function $L_i = \varphi(\tau)$, the operation regime of the furnace facilities for the considered time interval will be determined by the function $L_f = \omega(\tau)$ which is the difference $L_f = L_r - \varphi(\tau)$. Heat production of the furncae facilities for the considered period is determined as the integral $Q_f = \int_0^T \omega(\tau) d\tau$ and the value of their installed thermal capacity will correspond to the maximum value of $\omega(\tau)$ in this time interval $L_f^* = \omega_{max}(\tau)$, where $L_f^* \leq L_i^*$.

Analysis of actual regime of secondary energy resources output has shown that the schedule of secondary energy production for a given operating cycle can be considered to be invariable, only the amount of energy produced is subject to changes within some limits. The curve of secondary energy resources output in any other time interval (day, month, season) will be determined by the number of technological (operating) cycles which can occur within this time period. The most probable curve for a given technological aggregate can be statistically determined from sample data of the secondary energy output for a series of cycles in a year. These curves can be represented by the histograms in Fig. 4.2 to a sufficient degree of accuracy. For each time interval of a production process the value $L_j$ in each season is characterized by the thermal output of a source of secondary energy resources. The number of seasons $w$ and the length of each season should be proportional to the number of these time intervals and their length. Thus, to level off the curve of secondary energy output, the operation of the source of secondary energy resources during the cycle must be known. To determine the curve for another time interval the initial curve must be recalculated according to the number of cycles performed within the given time interval.

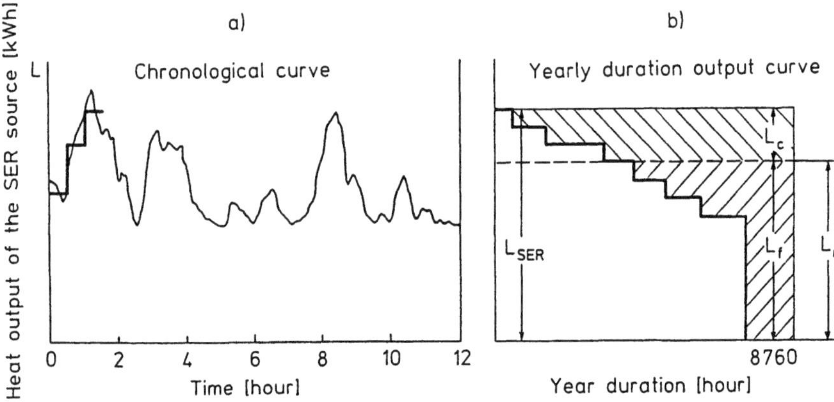


**Fig. 4.2.** Output curve of secondary energy resources by production aggregate: (*a*) – chronological curve, (*b*) – yearly duration output curve

The output of the furnace facilities in each season time interval can be determined as the difference in the output of the recovery installation and the source of secondary energy resources:

$$L_{f_j} = L_r - L_j$$

The highest value of $L_{f_j}$ is taken as the value of the installed furnace capacity $L_f^*$. Heat production by the furnace facilities for the considered time interval is determined as the sum of outputs in each season:

$$Q_f = \sum_{j=1}^{w} L_{f_j} \tau_j$$

The expression for the costs for the variant with utilization of secondary resources for a source $i$ becomes

$$Z_1' = \varepsilon_r K_r(L_r) + \varepsilon_f K_f(L_f^*) + a_f \beta_f Q_f + \varepsilon_c K k_c(L_c^*) \tag{4.7}$$

The costs of the second variant, without utilization of secondary energy resources of a source $i$, should obviously include only the costs of measures that would guarantee normal operating conditions of the technological aggregate itself and the gas cleaning system, i.e.,

$$Z_2 = \varepsilon_c K_c(L^*) . \tag{4.8}$$

Taking into account that the capacity of cooling devices is equal to $L_c^* = L^* - L_r$, the costs of utilizing secondary energy resources of a source $i$ can be written

$$\Delta Z = \varepsilon_r K_r(L_r) + \varepsilon_f K_f(L_f^*) + a_f \beta_f Q_f + \varepsilon_c K_c(L_r) . \tag{4.9}$$

### 4.2.2 Costs of Reliability Improvement

Estimation of the reliability is an important part of the analysis required to devise the optimal scheme of utilization of secondary energy resources [24, 42, 49, 73, 94, 141, 152, 194]. However, this problem is extremely complicated, since the recovery installation is the element that ties the regional energy supply system to an industrial plant into a united complex. Thus, it is necessary to consider the reliability of recovery installations not only as energy producing equipment, but also as an element in the chain of aggregates connected by the production scheme of a plant.

Failure of the heat recovery installation will lead to economic damage due to lower production of thermal, mechanical or electrical energy (depending on the specific configuration of secondary energy resources utilization) and to damage to the technology due to the disruption of the technological chain of aggregates. Determination of the total potential damage at a specific plant equipped with given technological aggregates with known reliability indices is a complex problem which is not considered here in detail. The influence of the reliablity of recovery installations on the efficiency of secondary energy resources utilization can be evaluated by the following procedure.

In general, taking into account reliability in optimizing their utilization is reduced to the search for the minimum expenditures using the expression

$$Z_1 = Z_1' + D_{\mathrm{tot}} = \min\,, \quad D_{\mathrm{tot}} = D_{\mathrm{t}} + D_{\mathrm{n}} \tag{4.10}$$

where $D_{\mathrm{t}}$ is the damage to the technology that is caused by the break in the chain of technological aggregates when the recovery installations fail; $D_{\mathrm{n}}$ is the national-economic damage caused by energy shortage.

As was noted above, determination of the total damage in specific cases is a complicated process and lies beyond the scope of this text. It can be shown, however, that $D_{\mathrm{tot}}$ for metallurgical, chemical and some other plants is much higher (in most cases by an order of magnitude or more) than the costs of installation of standby equipment. Therefore, it can be assumed that the damage caused by the failure of a heat recovery installation is inadmissible.

Such a statement must also take into account the additional costs of special measures to exclude any possible failures in the production process during emergencies of the recovery installations. Obviously, these costs must be much lower than the total damage, i.e., $Z_{\mathrm{tot}}^{\mathrm{add}} \leq D_{\mathrm{tot}}$. Otherwise, our arguments for a standby system become meaningless. Then expression (4.10) can be written in the form

$$Z_1 = Z_1' + Z_{\mathrm{tot}}^{\mathrm{add}} = \min \tag{4.11}$$

To determine $Z_{\mathrm{tot}}^{\mathrm{add}}$ it is necessary to consider the nature of each component of damage and hence, the measures needed to prevent it. These additional costs should be distributed among the measures to exclude the damage $D_{\mathrm{n}}$ due to heat (energy) shortage, on the one hand, and damage $D_{\mathrm{t}}$ due to a break in the technological chain, on the other hand. Redundancy in the heat recovery installations can be used as a simple measure eliminating $D_n$.

The methods applied for solving similar problems for common energy installations can also be used to determine the value of reasonable capacity reserves and hence, their costs $Z_{rs}^{add}$. However, it must be noted that the recovery installations are designed to work in conditions of high smoke content, higher corrosiveness of the cooled gases, irregular supply of gases, etc. and therefore, their reliability level is, as a rule, lower than for common energy installations.

The costs of redundancy are further determined by the following additional factors. As was noted earlier, utilization of secondary energy resources is intended for meeting basic heat loads of the region for industrial and communal needs. These heat consumers usually require an uninterrupted supply during the whole year. For example, the utilization time of the maximum heat loads for a complete cycle of a metallurgical plant accounts for about 8000 – 8500 hours. Usually, production aggregates which are sources of secondary energy resources cannot work for an entire year without shutdown for maintenance, since their operating period is considerably less than a year. Hence, recovery installations which use the heat of the waste gases of these aggregates also cannot produce thermal energy for an entire year. This must be taken into account in determining the economic efficiency of secondary energy resources utilization, i.e., the costs of their utilization must include some expenditures for standby energy installations. Otherwise, the total costs must include expenditures for large capacity reserves at central cogeneration plants or boilers. In each particular case both the method of redundancy and the associated expenses will differ.

Thus, in general, the damage component $D_n$ due to a heat (energy) shortage is substituted in (4.10) by $Z_{rs}^{add}$, which is determined with regard for the unique features of the specific equipment.

Additional costs on measures to eliminate $D_t$ can be made directly at the recovery installations to ensure their required reliability, or on installation of special devices to guarantee the normal operation of the main equipment (bypass flues, precleaning gas-cooling devices etc.) during failures of the recovery installations. The second option seems more advantageous since it is practically impossible to achieve the required reliability of recovery installations, particularly due to the noted difficulties in their operation. In this case, substituting $D_t$ instead of $Z_c^{add}$ in (4.11) we obtain

$$Z_1 = Z_1' + Z_{rs}^{add} + Z_c^{add} \tag{4.12}$$

In the preceding section, account was taken of the expenditures of construction and operation of bypass and cooling installations that are to ensure the given operating conditions of production equipment regardless of the capacity of the recovery installations. Their value was determined as a function of the difference in the maximum value of secondary energy resources output per unit time and the current value of capacity of the recovery installations (4.5): $Z_c = \varepsilon_c K_c(L_c^*)$ where $L_c^* = L^* - L_r$. Reliable operation of the recovery installations can be achieved by an increase in the capacity of bypass and cooling installations up to the total capacity of the recovery installations. This should guarantee normal operation of production equipment during breakdown of the heat recovery installations. This determines the costs of the cooling systems.

In such a statement of the problem (independent of the assumed capacity of the recovery installations) the complete cooling of the waste gases from the sources of secondary energy resources is achieved by additional costs equal to

$$Z_c^{add} = \varepsilon_c K_c(L_r) \tag{4.13}$$

Hence, the capacity of the cooling installations allowing reliable operation of recovery installations must equal the maximum capacity of the sources of secondary energy resources $L_c^* = L^*$ and the total expenditures are then equal to $Z_c = \varepsilon_c K_c(L^*)$, i.e., as in the case without secondary energy resources utilization.

Substituting these expressions for additional costs into (4.7) we get

$$Z_1 = \varepsilon_r K_r(L_r) + \varepsilon_f K_f(L_f^*) + a_f \beta_f Q_f + \varepsilon_c K_c(L_c^*) + Z_{rs}^{add} \, . \tag{4.14}$$

Comparing these costs to those arising for a system without secondary energy resources utilization (given by (4.8) and subtracting the secońd component of the costs from the first, we get

$$\Delta Z = \varepsilon_r K_r(L_r) + \varepsilon_f K_f(L_f^*) + a_f \beta_f Q_f + Z_{rs}^{add} \, . \tag{4.15}$$

For convenience, the dependence of the capital investment in energy installations on their capacity is usually approximated by $K = K_0 + kN$, where $K_0$ is fixed and $kN$ depends linearily on the capacity. Making the same approximation for the capital investment in recovery and furnace facilities, (4.15) can be written as

$$\Delta Z = \varepsilon_r \left(K_{0_r} + k_r L_r\right) + \varepsilon_f \left(K_{0_f} + k_f L_f^*\right) + a_f \beta_f Q_f + Z_{rs}^{add} \tag{4.16}$$

### 4.2.3 Calculation of the Minimized Total Costs

After the dependence of the costs of secondary energy resource utilization on the capacity of each secondary energy sources at the plant has been determined, we can begin to search for the overall optimized composition of the recovery system.

Suppose that there exists a plant with the number of sources of secondary energy resources equal to $n$. For each source, we know the maximum capacity $L_i^*$ of the secondary energy resources suitable for further utilization and the time dependence of their output $L_i = f(\tau)$. We need to find the composition of recovery installations at any value of $L_{SER} =$ const delivered by the plant to the energy supply system, at which the total costs of utilization are at a minimum.

Mathematically, this problem can be represented in the following way. We minimize the functional $\psi(L_{SER}) = \Sigma_{i=1}^{n} \Delta Z_i(L_{r_i}) \to \min$ subject to the following conditions:

$$0 \leq L_{\mathrm{T}} \leq L_i^* \; ; \quad \Sigma_{i=1}^n L_{\mathrm{T}_i} = L_{\mathrm{SER}} \; , \quad i = 1, 2, \ldots, n \; ; \quad \Delta Z_i(L_{\mathrm{T}_i}) \geq 0$$

The method for solving any problem is largely determined by the type of the functional and the limits on ranges of the variables or functions entering the functional. In our case it is easily seen that the function $\psi$ belongs to a class of separable functions, which are functions of $n$ variables that can be represented as a sum or a product of $n$ functions of one variable.

Problems involving separable functions with linear constraints can be solved by "dynamic programming" methods [19, 20, 193]. The principal idea of dynamic programming is that rather than optimizing functions of many variables, one optimizes, step-by-step, functions of only two variables. This is achieved by reducing the problem to the solution of certain recurrent functional equations. Such equations contain both ordinary variables for the system state and a group of control parameters serving as a feedback. The proper variation at each step of the process should lead to the most effective way of achiving the goal.

Thus, dynamic programming is particularly suited for multi-step analysis of continuous processes and allows solution of complex engineering problems of high dimensionality.

In this section we demonstrate the method to the case of optimizing the composition of recovery installations. This is a particular case of *Bellman*'s optimality [19, 20, 193] and can be solved by a functional equation of a general type:

$$\begin{aligned} &\psi(L_{\mathrm{SER}}) = \min \left[\varphi_1 \left(L_{\mathrm{T}_i}\right) + f_{n-1} \left(L_{\mathrm{SER}} - L_{\mathrm{T}_i}\right)\right] \\ &0 \leq L_{\mathrm{T}_i} \leq L_1^* \\ &0 \leq L_{\mathrm{SER}} \leq \Sigma L_i^* \; , \quad \varphi_1 \left(L_{\mathrm{T}_i}\right) = \Delta Z_1 \left(L_{\mathrm{T}_1}\right) \; . \end{aligned} \tag{4.17}$$

The optimal contribution of a recovery installation is governed by the particular scheme of the multi-step process. Initially, the contribution of a pair of recovery installations participating in the heat production of a given volume $L_{\mathrm{SER}}$ ranging between the limits $0 \leq L_{\mathrm{SER}} \leq L_1^* + L_2^*$ is optimized ($L_1^*$ and $L_2^*$ are the maximum possible capacities of sources 1 and 2, respectively). In this case the functional equation for the optimality has the form

$$\begin{aligned} &\psi_{1,2}(L_{\mathrm{SER}}) = \min \left[\varphi_1 \left(L_{\mathrm{T}_2}\right) + f_2 \left(L_{\mathrm{SER}} - L_{\mathrm{T}_2}\right)\right] \\ &0 \leq L_{\mathrm{T}_2} \leq L_2^* \; . \end{aligned} \tag{4.18}$$

Then, the minimum of the functions $\psi_{1,2}$ and $f_3$ of three recovery installations for different values of $L_{\mathrm{SER}}$ ranging between $0 \leq L_{\mathrm{SER}} \leq L_1^* + L_2^* + L_3^*$ is calculated. The functional equation for this case is

$$\begin{aligned} &\psi_{1,2,3}(L_{\mathrm{SER}}) = \min \left[\psi_{1,2}(L_{\mathrm{SER}}) + f_3 \left(L_{\mathrm{SER}} - L_{\mathrm{T}_3}\right)\right] \\ &0 \leq L_{\mathrm{T}_3} \leq L_3^* \; . \end{aligned} \tag{4.19}$$

The procedure is repeated for the $n$ recovery installations. The overall minimum costs of secondary energy resources utilization for any assortment of installations is determined by successive evaluation of th partial optima.

Figure 4.3 presents a flow chart of this algorithm. Let the current value of the heat output of the recovery installations at a plant be $L_{SER}$ and assume that this volume of $L_{SER}$ is generated by the source 1. Calculate the costs to produce this volume. Since in this case there are no other alternatives for heat production, these are the minimum costs. By varying the value of $L_{SER}$ (which should not exceed the maximum value of heat produced by the source) and calculating the costs, we construct a table for $Z_{pr} = f(L_{SER})$. In the next step it is assumed that two sources, 1 and 2, participate in heat production. We set $L_{SER}$ to be $\Delta L$ away from the lowest value and find the effect of source 2 on the total expenditures. For this purpose the contribution of source 2 is given as $x$ at the fixed value of $L_{SER}$ and the costs for it are determined. The remaining part of thermal load $(L_{SER} - x)$ must be met by source 1, therefore, from the table for $Z_{pr} = f(L_{SER})$ the value of the costs $Z_{pr}$ corresponding to $(L_{SER} - x)$ is chosen and added to the costs for source 2. Thus, the expenditures of the plant to produce $L_{SER}$ by two sources are

$$Z_{tot} = Z_2 + Z_{pr} \, .$$

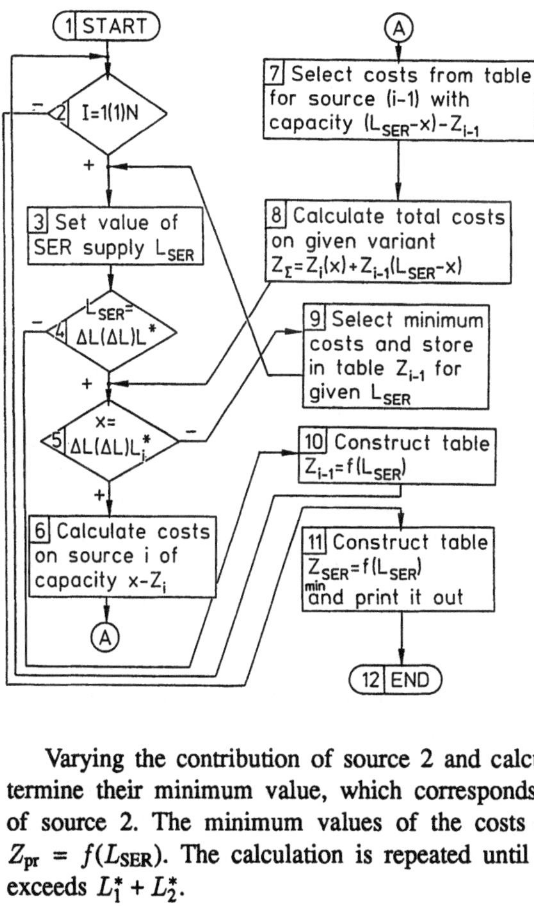


**Fig. 4.3.** Flow chart of the program to find characteristics for heat recovery installations

Varying the contribution of source 2 and calculating $Z_{tot}$ each time, we determine their minimum value, which corresponds to the optimum contribution of source 2. The minimum values of the costs are included in the table for $Z_{pr} = f(L_{SER})$. The calculation is repeated until the volume of heat produced exceeds $L_1^* + L_2^*$.

The calculation is repeated for source 3, and so on. Note that the sum of all the sources already optimized is taken as the preceding source, and the minimized costs calculated from the optimized combination of sources to produce each current value of $L_{\text{SER}}$ are taken for $Z_{\text{pr}}$. The value of $x$ must not exceed the maximum possible volume of the heat output $L_i^*$ of source $i$ and $(L_{\text{SER}} - x)$ may not exceed the sum of the maxima of all the preceding sources.

Thus, when $L_{\text{SER}} = \Sigma_{i=1}^{n} L_i$ is found the minimized costs to generate any amount of secondary energy resources to the energy supply system are given in the table. The relationship $Z_{\text{SER}} = \psi(L_{\text{SER}})$ represents an "equivalent characteristic" of recovery installations at a plant.

Each point of such a characteristic is associated with the optimal internal parameters of the optimized subsystems: the most favorable combination and output of the recovery installations which generate a given amount of power to the energy supply system of the region, i.e.,

$$\underset{\min}{Z}\,_{\text{SER}} = f(L_i)\,, \quad i = 1, 2, \ldots, n\,.$$

The equivalent characteristic is, as a rule, a concave curve, since an increase in the use of secondary energy resources is achieved by utilization of that particular part that requires more complex designs and larger material expenditures. Similar characteristics can be determined by the technique described here for other industrial plants.

## 4.3 Determination of the Optimal Extent of Secondary Energy Resources Utilization at an Industrial Plant

As we have shown in the previous sections, the optimum extent of utilizing available secondary energy resources can be found by searching for the most favorable structure of generating capacities in the energy supply system using recovery installations for meeting thermal and electrical loads in a given region. The algorithm for solving this problem contains elements of the energy supply scheme of the plant, the features of the ties between the energy supply system and the interconnected power system and on the route of secondary energy resources use. We will discuss this algorithm for the most widely used thermal route.

Large industrial plants usually have their own or adjacent cogeneration plants which meet both the thermal loads of neighboring consumers and electrical loads of the local power system which serves the industrial region. The heat recovery installations, therefore, compete with the cogeneration plant to meet the thermal load curve.

To find the optimum extent of secondary energy resources utilization in already existing antiquated plants relying on only cogeneration facilities it is necessary to understand that utilization of heat produced by added recovery installations will deteriorate the operating conditions of the cogeneration plants since it will reduce the amount of heat extracted from them. That is, the addition of heat recovery installation will decrease the electricity production of the cogeneration cycle and increase it in a condensation cycle. This, in turn, will lead to fuel overfiring at the cogeneration plant, which can be reduced by decreasing the electricity production in the condensation cycle through participation of the cogeneration plant in meeting peak load of the power system. If the cogeneration plant has back-pressure turbines then a decrease in the extracted heat due to the contribution of the heat recovery installations can lead to disconnection of some power units. The resulting power shortage can be offset by the back-up capacity of the condensation power plants. Otherwise, the presence of cogeneration plants in the system reduces the efficiency of secondary energy resources utilization and decreases the contribution of recovery installations of the industrial plant in covering thermal loads of the region. However, these negative effects of their utilization can be treated as temporary. Further growth of heat consumption in subsequent years will normalize the operation of the cogeneration plant.

Optimization of the scheme of secondary energy resources utilization for a designed plant can be considered as a problem of future planning. The inclusion of SER can lead to changes in the designed ratio of heat production by a cogeneration plant to the total thermal load, i.e., in the decrease of cogeneration plant capacity and hence, to a change in the structure of generated powers and heat capacities in the energy supply system of the region.

Depending on the ratio between the thermal loads of the region and the heat output of recovery installations it may turn out to be undesirable to install back-pressure turbines at the cogeneration plants or to devise an energy supply scheme without these plants. The methods for solving the stated problem will be presented below for the conditions of future design.

The thermal loads of a region are met by cogeneration plants through the turbine extractions, industrial boilers and peak boilers, thus, the recovery installations can be considered as a source to help meet the load curve. The available sources must participate in heat generation for the consumers of the region at the minimum total discounted costs, i.e., this system should be considered as a single complex. In the complex optimization, the optimal parameters of each component are determined by the minimum total costs on output of a given volume of product by all the related branches.

The thermal and electric loads of a region that are to be covered at minimum cost may be considered as a given volume of product of the energy complex. The efficiency of using secondary energy resources should be determined from the minimum costs for the whole complex:

$$\Sigma Z_{\text{compl}} = \min$$

We will optimize the utilization of secondary energy resources by constructing a mathematical model on the basis of decomposition, allowing to solve in each block the individual problems. We assume that only some fraction of the available heat capacity equal to the sum of the maximum values of the capacity of all the existing sources of secondary energy resources ($L^*_{\mathrm{SER}} = \Sigma^n_{i=1} L^*_i$) is used. In general, this value is obtained by sets of recovery installations differing in composition and output and is described by the vector $\boldsymbol{L}_i$. Participation of secondary energy resources in meeting thermal load curves changes the optimum structure of the generating powers and heat capacities of the energy supply system of the region the operating conditions of energy equipment and hence, leads to a change in the minimum discounted costs of the energy supply system. Thus, we have the relationship

$$\underset{\min}{Z}_{\mathrm{ESS}} = f_1(\boldsymbol{L}_i)\,, \qquad i = 1, 2, \ldots$$

As was shown above, the costs of secondary energy resource utilization can be represented, in the end as a function of the same variables

$$\underset{\min}{Z}_{\mathrm{SER}} = f_2(\boldsymbol{L}_i)\,, \qquad i = 1, 2, \ldots\,.$$

The optimal parameters of the components of the complex can be determined subject to

$$\Sigma Z_{\mathrm{compl}} = \underset{\min}{Z}_{\mathrm{ESS}} + \underset{\min}{Z}_{\mathrm{SER}} \rightarrow \min \tag{4.20}$$

The representation of the costs of the complex as a sum of components depending on the same variables allows us to separate the model into blocks and solve the minimization problem of the total costs $Z_{\mathrm{compl}}$ of the complex in parts. We only need to define the parameters of the connections between the blocks. Block 1 minimizes the costs of secondary energy resources utilization taking into account specific conditions of work for each source and their route of utilization for each current value of the plant output $L_{\mathrm{SER}}$ that can be generated so the energy supply system of the region. Block 2 determines the optimal future structure of the generating capacities of the energy supply system, if the corresponding output $L_{\mathrm{SER}}$ of the plant is included in the considered sources. Block 3 relates the results obtained in the preceding blocks, determines the minimum costs of the complex and the corresponding optimal parameters of the energy supply system and $L^{\mathrm{opt}}_{\mathrm{SER}}$.

Figure 4.4 presents this scheme. The problem is formulated as follows. We have a hypothetical plant with a given technology and hence a given volume and curve of output of secondary energy resources $L^*_{\mathrm{SER}}$ and $L_{\mathrm{SER}} = f(\tau)$. This plant in the region is designed with known thermal and electrical loads. The capacity of the existing condensation power plants $N_{\mathrm{ec}}$ is given. It is assumed that the power system consists of thermal power plants only. We determine the optimal structure of the energy supply system, including the contribution of recovery installations of the industrial plant, to cover the thermal load of the region. Or, it is necessary to find the most advantageous variant for meeting a growth in the thermal and electrical loads of the region by transmission of power from the interconnected power system, construction of new condensation and cogeneration plants, industrial boilers, peak boilers or recovery installations.

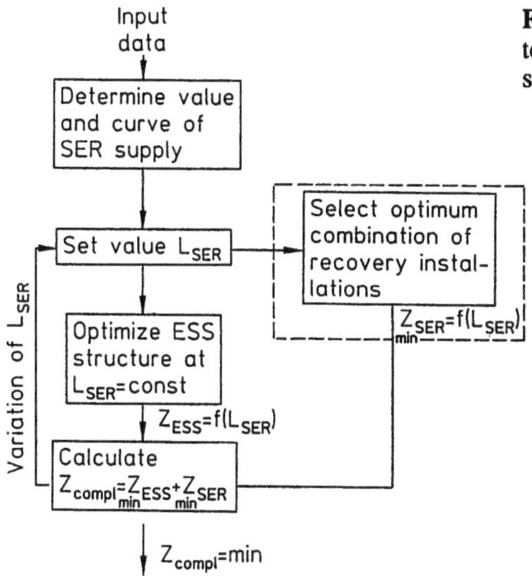


**Fig. 4.4.** Scheme for optimizing the extent of utilization of secondary energy resources

Thus, the initial information includes: future electrical and thermal load curves, the capacity of existing thermal power plants, "techno-economic" indices of power plants, peak boilers, industrial boilers etc. The anual heat consumption is represented by a histogram with $w$ periods (seasons of the year) of duration $\tau_j$ and in each of them the value of thermal load is constant. The corresponding electrical load duration curves having both peak and base parts are then constructed for these seasons.

In applying these methods to problems of future design, the operating conditions of particular energy installations can be optimized in a simplified way according to the values of their specific fuel consumption. The contribution of each type of power plant in meeting electrical demand is taken in the order of increasing consumption. The base part of the curve is met by cogeneration plants having the lowest fuel consumption per kilowatt–hour and then by new highly efficient condensation power plants (transmission from the interconnected power system), then by existing less energy efficiency power plants and finally by cogeneration plants operated in a condensing mode (Fig. 4.5).

The distribution of contributions of energy installations participating in meeting a regional thermal load curve is similar output of recovery installations to meet the base load, then thermal energy from turbine extractions of cogeneration plants and finally production of industrial boiler and peak boilers (Fig. 4.6).

The optimal structure of the energy supply complex is determined by three parameters: the heat output of utilized secondary energy resources $L_{SER}$, the electric capacity of cogeneration plants $N_{cp}$ and the output of industrial boilers $L_{ib}$. They can be determined by minimizing the total discounted costs of the energy supply system of the area:

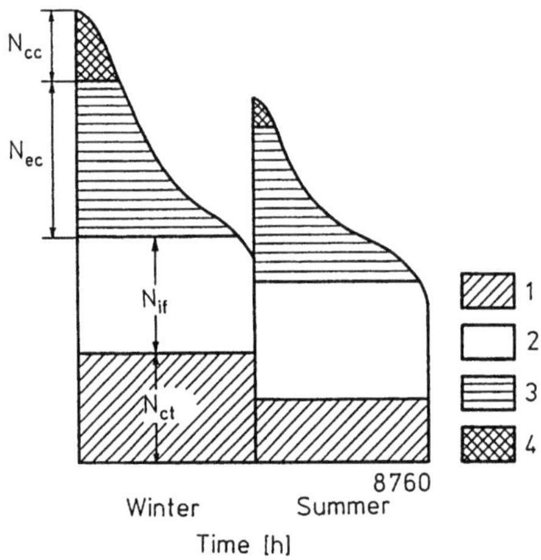


**Fig. 4.5.** Contribution of power plants in meeting an electrical load curve, thermal secondary energy resources. 1 – cogeneration plant operating in thermal mode; 2 – intersystem power flow; 3 – existing condensation power plants; 4 – cogeneration plant operating in a condensing mode

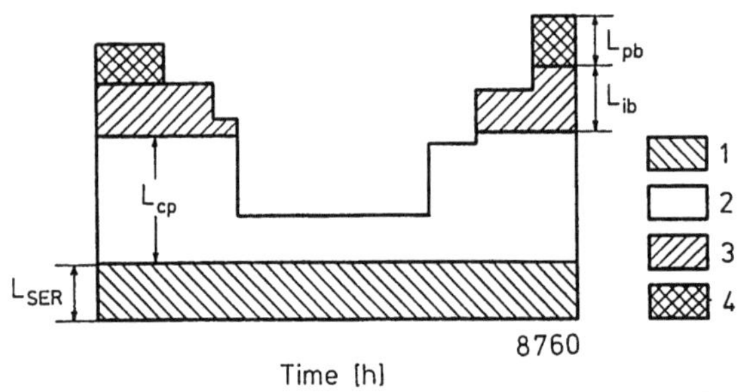


**Fig. 4.6.** Contribution of energy installations in meeting a thermal load curve. 1 – heat recovery installations; 2 – cogeneration plants; 3 – industrial boilers; 4 – peak boilers

$$
\begin{aligned}
Z_{\mathrm{ESS}} &= \varepsilon_{\mathrm{cp}}\left(K_{0_{\mathrm{cp}}} + k_{\mathrm{cp}}\right) N_{\mathrm{cp}} + z_{\mathrm{cp}}\left(\beta_{\mathrm{ct}} \Sigma_{j=1}^{w} W_{\mathrm{ct}_j} + \beta_{\mathrm{cc}} \Sigma_{j=1}^{w} W_{\mathrm{cc}_j} + \beta_{\mathrm{cp}}^{\mathrm{h}} \Sigma_{j=1}^{w} Q_{\mathrm{cp}_j}\right) \\
&+ z_{\mathrm{ec}} \beta_{\mathrm{ec}} \Sigma_{j=1}^{w} W_{\mathrm{ec}_j} + \varepsilon_{\mathrm{ib}}\left(K_{0_{\mathrm{ib}}} + k_{\mathrm{ib}} L_{\mathrm{ib}}\right) + z_{\mathrm{ib}} \beta_{\mathrm{ib}} \Sigma_{j=1}^{w} Q_{\mathrm{ib}_j} \\
&+ \varepsilon_{\mathrm{pb}}\left(K_{0_{\mathrm{pb}}} + k_{\mathrm{pb}} L_{\mathrm{pb}}\right) + z_{\mathrm{pb}} \beta_{\mathrm{pb}} \Sigma_{j=1}^{w} Q_{\mathrm{pb}_j} \\
&+ Z_{\mathrm{hnw}} + z_{\mathrm{el}} \Sigma_{j=1}^{w} W_{\mathrm{if}_j} + Z_{\mathrm{enw}} + Z_{\mathrm{SER}} = \min
\end{aligned}
\tag{4.21}
$$

where $K_{0_{\mathrm{cp}}}$, $K_{0_{\mathrm{ib}}}$, $K_{0_{\mathrm{pb}}}$ are the fixed components of the investment costs of cogeneration plants, industrial boilers and peak boilers, respectively; $k_{\mathrm{cp}}$, $k_{\mathrm{ib}}$, $k_{\mathrm{pb}}$ are additional specific capital investments in these installations; $\varepsilon_{\mathrm{cp}}$, $\varepsilon_{\mathrm{ib}}$, $\varepsilon_{\mathrm{pb}}$ are the sum of investment cahrges due to depreciation, routine repair and servicing of these installations; $z_{\mathrm{cp}}$, $z_{\mathrm{ec}}$, $z_{\mathrm{ib}}$, $z_{\mathrm{pb}}$ are the prices of fuel burnt at cogeneration plants, existing condensing power plants, industrial boilers and peak boilers, respectively; $\beta_{\mathrm{ct}}$, $\beta_{\mathrm{cc}}$, $\beta_{\mathrm{ec}}$ are the specific fuel consumption to generate 1 kWh of electrical energy by thermal and condensing modes at cogeneration plants and at existing condensation power plants; $\beta_{\mathrm{ct}}^{\mathrm{h}}$, $\beta_{\mathrm{ib}}$, $\beta_{\mathrm{pb}}$ are the specific fuel consumption to generate 1 GJ of heat at cogeneration plants, industrial boilers and peak boilers; $W_{\mathrm{ct}_j}$, $W_{\mathrm{cc}_j}$, $W_{\mathrm{ec}_j}$ are the levels of electricity production at cogeneration plants by thermal and condensing modes and at existing condensation power plants in a season $j$; $W_{\mathrm{if}_j}$ is the power transmitted from the interconnected power system in a season $j$; $Q_{\mathrm{cp}_j}$; $Q_{\mathrm{ib}_j}$, $Q_{\mathrm{pb}_j}$ heat output at cogeneration plants, industrial boilers and peak boilers in a season $j$; $z_{\mathrm{el}}$ and the prices of electrical energy in the considered region; $Z_{\mathrm{hnw}}$, $Z_{\mathrm{enw}}$ are the costs of heat and electrical networks; $Z_{\mathrm{SER}}$ are the costs of secondary energy resources utilization.

For an isolated or nearly isolated local power system (which happens rather rarely) the power and energy shortage resulting by the replacement of cogeneration plants with recovery installations is offset by the increase in capacity of the new condensation power plants. Therefore, instead of $z_{\mathrm{el}} \Sigma_{j=1}^{w} W_{\mathrm{if}_j}$, the total costs should include the costs of the additional power $N_{\mathrm{nc}}$ and the output $W_{\mathrm{nc}}$ of the new condensation power plants

$$\varepsilon_{\mathrm{nc}} k_{\mathrm{nc}} N_{\mathrm{nc}} + z_{\mathrm{nc}} \beta_{\mathrm{nc}} \Sigma_{j=1}^{w} W_{\mathrm{nc}_j} \ .$$

In principle the approach to the problem of choosing the optimal degree of utilization of secondary energy resources in the combined route is the same as for the thermal one. The same assumptions can be made: the power system consists of thermal power plants; heat loads are covered by extractions of cogeneration plants, industrial boilers, peak boilers and extractions of heat recovery cogeneration plants.

For the combined regime, as for the thermal one, the maximum value $L_{\mathrm{SER}}^{*}$ and the output curve of secondary energy resources $L_{\mathrm{SER}} = f(\tau)$, the existing state of the system, techno-economic indices of installations are given. It is necessary to determine the most advantageous variant for meeting the growing thermal and electrical loads of the region, when the heat recovery cogeneration plant is used as a source of electricity and heat.

The operating conditions of generating plants are chosen according to the fuel consumption in energy production. The base part of electric load curves is met primarily by cogeneration capacity of heat recovery and local cogeneration plants. Then power from the interconnected power system or new condensation power plants is used. These are followed by the existing condensation power plants and lastly, by the condensation power of cogeneration plants (Fig. 4.7).

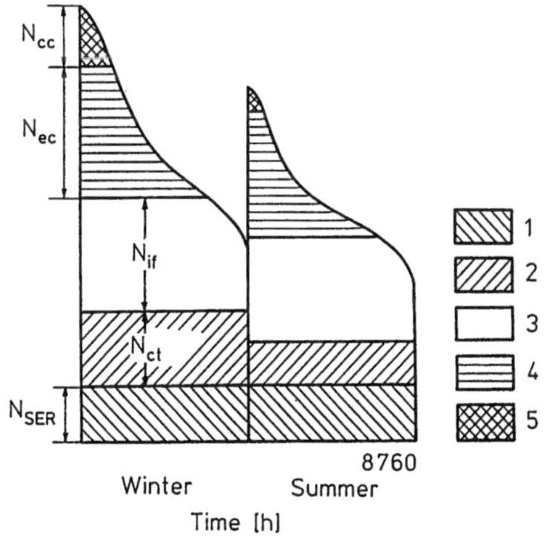


**Fig. 4.7.** Contribution of power plants in meeting an electrical load curve in a combined route of secondary energy resources utilization. 1 – heat recovery installations; 2 – cogeneration plant operating in the thermal mode; 3 – intersystem power flow; 4 – existing condensation power plants; 5 – cogeneration plant operating in condensing mode

The thermal load curve, as in the case of the thermal route of secondary energy resources utilization, is first of all covered by extractions of heat recovery cogeneration plant, then by cogeneration extractions of the local cogeneration plant, industrial boilers, and peak boilers (Fig. 4.6).

The heat output of the utilized secondary energy resources and the composition of the electrical and heat energy sources in the energy supply system are the parameters to be optimized. In the thermal regime the heat produced in the recovery installations $L_{\mathrm{SER}}$ is completely used for heat supply, whereas in the combined regime it is supplied as steam into the cogeneration turbine. In this case the secondary energy resources are used to generate the electrical $N_{\mathrm{SER}} = L_{\mathrm{SER}}/q_{\mathrm{n}}$ and heat $L_{\mathrm{SER}} = N_{\mathrm{SER}}/e_{\mathrm{q}}$ energy where $q_{\mathrm{n}}$ is the specific heat consumption for production of 1 kWh of electrical energy for a given amount of steam extraction; $e_{\mathrm{q}}$ is the specific amount of electrical energy produced by the heat recovery cogeneration plant per unit of heat consumption.

In seasons in which the capacity of heat extracted from the heat recovery cogeneration plant exceeds the thermal load, the plant can work partially in the condensing cycle:

$$N^{\mathrm{c}}_{\mathrm{SER}_j} = \left(L_{\mathrm{SER}} - L_{\mathrm{ld}_j}\right)/q_{\mathrm{c}} \tag{4.22}$$

where $L_{\mathrm{ld}_j}$ is the thermal load in a season $j$; $q_{\mathrm{c}}$ is the specific heat consumption for production of 1 kWh of electrical energy by the heat recovery cogeneration plant in the condensing cycle. The rest of the problem can be solved by using the same algorithm as that described above for the thermal route of secondary energy resources utilization.

# 5. Energy Balances in Ferrous Metallurgy

Consumption of iron and its alloys currently accounts for nearly 90 % of the total consumption of all metals. This fact above all stipulates the urgency of increasing the energy efficiency of ferrous metallurgy, as even a small increase can have a net considerable effect on the sector on the whole.

Iron ores belong to the group of oxidized ores. Iron is mainly present in the form of hematite ($Fe_2O_3$) and magnetite ($Fe_3O_4$). Before transport to metallurgical plants the mined ore is usually concentrated. It is crushed, sorted according to the size of the lumps, washed to remove sand and clay and then magnetically concentrated. Iron ore concentrate contains up to 65 % Fe.

Ferrous metals – iron and steels of different grades – are produced by pyrometallurgical methods. Blast furnace processing is currently, and will continue to be in the near future, the most common method of iron processing and steel is produced in open-hearth and electric furnaces and oxygen converter.

In the USSR, further development in iron production will be aimed at increasing the volume of the blast furnaces up to 6000 $m^3$, intensifying the blast furnace efficiency by using the oxygen-enriched air, reducing the consumption of deficient and expensive coke by substitution of gas and fuel oil. Improvement in steelmaking suggests increases in the capacity of the oxygen converter and electrical furnace shops, and gradual reduction in use of the open-hearth stage.

Nevertheless, the open-hearth process will be important in steel production until at least the end of the twentieth century. For this reason, in this chapter we take as our example a metallurgical complex with an open-hearth to illustrate the technique of exergy analysis of technological processes described in Chaps. 2 and 4.

## 5.1 The Production Scheme

Consider a complex with complete metallurgical and coke and coking by-product cycles. Along with the main technological shops, the complex has an energy supply service that includes a cogeneration plant, heat networks, electrical services, an oxygen compressor plant, a water supply and sewage. In addition, the complex has a number of auxiliary shops: engineering, forge, shaped casting and shops for maintenance of furnaces, refractories, etc.

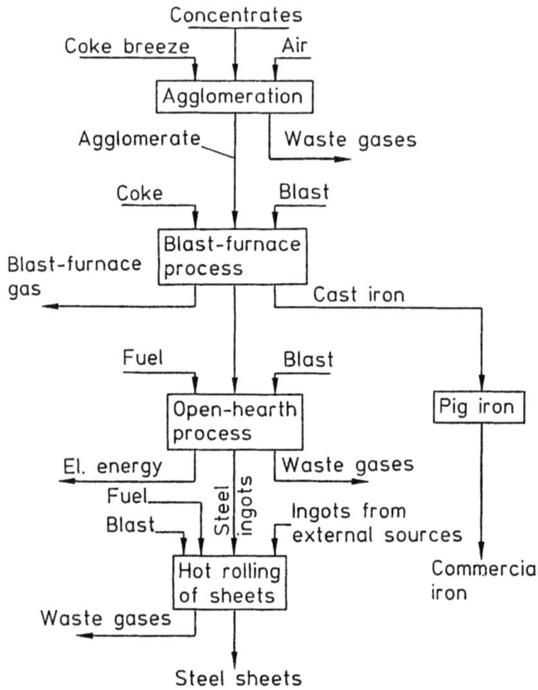


**Fig. 5.1.** Metallurgical Production Scheme

### 5.1.1 Metallurgical Cycle

According to the scheme given in Fig. 5.1, the metallurgical cycle includes an agglomerating plant, open-hearth and blast-furnace shops, and a rolling mill. Let us briefly discuss each of them.

**Agglomeration process.** The operation of blast furnaces is more efficient when the lumps of raw material are not brittle and have a high iron content. As mentioned above, to increase the iron content the ore is preliminarily crushed. Agglomeration and pelletizing of the crushed ore allows lumps of reasonable size to be produced.

The sintering oven is charged with iron-containing concentrates, dust of sintered wastes and ore and coke. The charge passes through the oven to a hearth. Here, the coke ignites.

The primary processes occurring in the oven are sulphur oxidation, water evaporation and silicate formation. In addition ferrous silicate, with a relatively low melting point of 1209°C, is formed via

$$2Fe_3O_4 + 3SiO_2 + 2CO = 3Fe_2SiO_4 + 2CO_2 \ .$$

The charge is heated to 1300°C so that the lumps melt together. During cooling they form a porous mass, i.e., agglomerate.

The cooled agglomerate chunks are forwarded to the sizing screens, where they are sorted according to size. Fine lumps are returned to the sintering ovens and large lumps (25 mm dia.) are fed into the blast furnaces.

**The Blast-furnace process.** The charge of the blast-furnace smelter contains the agglomerate, fluxes, other iron-containing materials and coke. To reduce the consumption of coke, coal dust and fuel oil are also blasted into the furnace. The blast air is heated in Cowper hot-blast stoves to 1000 – 1200°C using a mixture of blast-furnace and coke-oven gases.

The iron oxides are reduced at a temperature of 500 – 1100°C. A so-called indirect reduction takes place at the lower temperatures in this range, with CO-containing gases being the reducing agents. Direct reduction occurs at higher temperatures, when soot carbon is the reducing agent. Forty to sixty percent of the iron is reduced directly. Iron carbonization takes place above 1200°C:

$$3Fe + 2CO = Fe_3C + CO_2 ,$$

$$3Fe + C = Fe_3C .$$

Reduction of manganese, silicon and phosphoric oxides proceeds in parallel with that of the ferric oxides.

Sulphur present in the furnace in the coke, the ore, and limestone considerably decreases the iron grade. It can be removed by inserting slags with high CaO content.

The blast-furnace process yields iron and slag with a temperature of about 1500°C, blast-furnace gas with a temperature of 250 – 300°C and a pressure of $2.1 - 2.5 \times 10^5$Pa.

The blast-furnace gas is combustible; its low calorific value is $Q_1^{rec} = 4000 kJ/m^3$ (under standard conditions), therefore, it is cleaned from dust and used as fuel. Slag from blast furnaces is dumped. Part of the iron produced in the blast furnaces is poured into the open-hearth furnaces and the remaining iron is fed to the casting machines.

**The Open-hearth process.** The open-hearth process is used to produce steel of different grades out of pig iron. Flame-contact furnaces are charged with a mixture of liquid iron and solid materials – iron and steel scrap, iron ore, limestone, etc.

Smelting in an open-hearth furnace occurs thermally by the heat released in fuel combustion and also chemically in the oxidation reactions of carbon, silicon and manganese that are contained in the iron.

Silicon and manganese oxides yielded by these reactions pass into the slag. In addition, some of the iron is re-oxidized: $2Fe + O_2 = 2FeO$. But the ferric monoxide produced by this reaction does not pass into slag, as it is re-reduced by iron impurities that have not been oxidized.

These secondary reactions play a decisive role in the steelmaking process, as they decrease the metal losses. The regenerator is an important component of the open-hearth furnace. The air in them is heated to 1000 – 1200°C (if the furnaces are fueled by mixed or generator gas, it is also heated in regenerators).

**Rolling.** Ingots produced in the steelmaking shops are sent to rolling shops for further processing. Rolling requires several stages. In the first stage the ingots are rolled in the primary mills to bloom or slab. Slabbing mills transform ingots into slabs for rolling mills. The primary mill is equipped with pit furnaces to heat the ingots to the required temperature of about 1200°C. The pit furnaces are fueled by either a mixture of coke-oven and blast-furnace gases, or by fuel oil. The operating temperature of the pit furnaces is about 1350 – 1400°C. They are equipped with ceramic recuperators which heat the blast air to 800 – 850°C.

Different section or rolling mills are usually installed after the primary mills to roll the slabs into bands of different thickness and width. They are usually equipped with continuous furnaces for heating the slabs to the rolling temperature. In the gas ducts of furnaces there are ceramic recuperators, where the air fed into the furnace is heated to 750 – 950°C.

### 5.1.2 Coking By-product Cycle

Metallurgical complexes usually include a coke and coking by-product cycle and a cogeneration plant. This combination is economical because the energy supply becomes much cheaper. Coke-oven and blast-furnace gases and secondary energy resources are available and the volume of goods requiring transportation decreases.

The coke and coking by-product processes include coal preparation plants, coke-oven batteries, a facility for coke-oven gas purification and extraction of valuable components from it. The basic scheme is shown in Fig. 5.2.

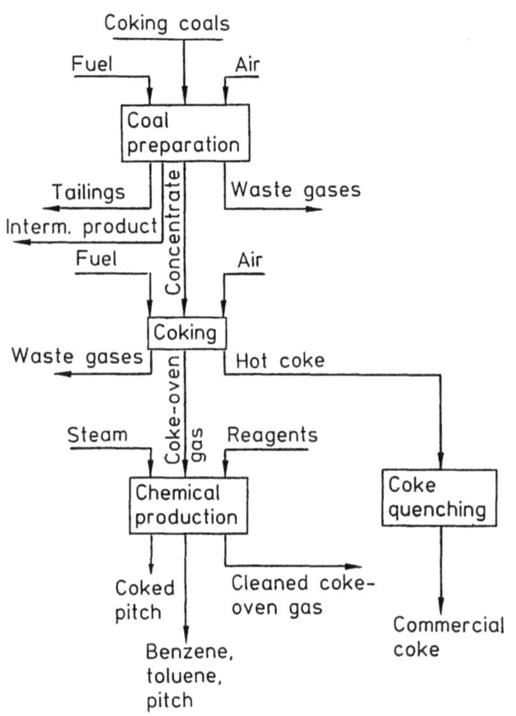


**Fig. 5.2.** Scheme of coke and coking by-product production

**Coal preparation.** To produce coke with an ash content of < 13 %, as is required in metallurgy, the ash content of the initial charge should be less than 10 %. Coking coals have, as a rule, a higher ash content than this and must therefore be preprocessed.

The coal preparation process includes a coal crushing stage, a fractionation, preparation of the large-size fractions in heavy media, the mid-size fractions via sedimentation and the small-size ones – by flotation, and concentrate drying.

The preparation process yields the concentrate needed for coking with an ash content of $\leq$ 10 % and moisture content of about 8 %. In addition, intermediate products, slime and tailings are produced.

**Coking.** Coking involves relatively slow heating of a charge of coking coals to a temperature of about 1100°C in a special air-tight chamber. When the coal softens, it releases practically all the volatile components, decreasing their content in coal from $V_{\text{ave}}^{\text{vol}} = 26 - 28\,\%$ to $V_{\text{ave}}^{\text{vol}} = 0.8\,\%$ in coke. This so-called coke-oven gas is a product of the reaction of coke and its volatile components with water vapor, hot chamber walls, etc. The temperature of the gas leaving the chamber varies from 200 to 800°C. Afterwards it is cooled, cleaned and can be used as a fuel. For example, a mixture of the coke-oven and blast-furnace gases is used for heating the coke-ovens. The calorific value of the coke-oven gas is rather high, 19000 – 20000 kJ/m$^3$.

After a 14-hour heating only the porous, non-brittle blast-furnace coke remains in the chamber. Typically it is at a temperature of 1000 – 1100°C when it is sent on for dry or wet (water) quenching. Dry quenching has lately been widely used. Coke dry-quenching plants allow the heat from the hot coke to be used for steam generation.

**Chemical production.** All the steps related to the retrieval of valuable (and toxic) products from the coke-oven gas belong to the chemical production.

Containment, separation, purification, which are involved in the coke-oven gas cleaning, all consist in the successive heating and cooling stages. For example in coke-oven gas purification it is first cooled with water to condense out the major part of the water vapor and tar. The processes of purification and distillation consume the amount of heat needed to evaporate of the final product.

## 5.2 Energy Balances of the Metallurgical Complex and its Main Shops

Energy balances of our sample metallurgical complex and its main shops were drawn using the technique described in Chap. 2 and compiled in the tables discussed below. In the first column of the tables are the balance items which are the values of materials, energy, products and semiproducts consumed per unit of primary final product of the complex, i.e., per ton of rolled steel. The kilowatt-hour is taken as the measurement unit of different energy forms.

Table 5.1 presents the energy balance of the metallurgical complex calculated only from input and output parameters without account of internal transformations and flows of energy. The input part of the balance shows the volumes and structure of energy consumption and the output part the energy and exergy of final products and total energy losses.

The only external inputs are chemical energy in the form of coal, fuel oil, metal, tar and various iron-containing raw materials. All the supplied energy is eventually converted into chemical energy of the final products (rolled steel, pig iron, coke and coking by-products, etc.) as well as into heat.

The values of chemical energy and exergy for each material are determined by using the techniques described in Chap. 2 and are also given in the tables. Because the materials from external sources and products for external consumers have the temperature of the environment, their heat energy and exergy are equal to zero.

As is seen from the table, the total energy and exergy of the useful products are 10.238 MWh and 9.256 MWh and the consumption for their output is 26.427 MWh and 25.28 MWH, respectively. The ratios of these values yield the total energy and exergy efficiencies of the complex equal to 38.8 % and 36.6 %. The total energy losses (16.189 MWh) and exergy losses (16.024 MWh) are determined by the difference. It should be noted that these figures also include energy consumption in the auxiliary shops, for start-up and adjustment services.

The balance shows, that the ratio of energy carriers for the considered production process is

$$I_r : I_f : E_{el} : I_q = 0.09 : 0.91 : 0.0 : 0.0 ,$$

where the value of the chemical energy $I_r$ implies the chemical energy of ores, raw material and other non-combustibles. This value is the consumption of non-conventional energy carriers that are generally not included in predictions of the necessary amount of energy carriers.

The input part of the balance does not include electricity and heat, which are certainly needed both in the production processes and for auxiliary services of the complex. This is because the electricity and heat demand is met by the cogeneration plant of the complex which mainly operates on the intermediate product and blast-furnace gas. The cogeneration plant completely meets the plant's requirements for electricity and heat and supplies external consumers with the excess. This is shown in the output part of the balance.

This energy balance does not allow determination of the structure and source of losses and, hence, evaluation of the possibilities of improving the energy use. These data can be obtained from the more detailed energy balances calculated for individual technological processes or shops.

Tables 5.2–9 present the energy balances of the main shops with their auxiliary services. These balances are calculated in a similar way as the balance of the entire complex, i.e., each shop is considered as an independent thermodynamic system

**Table 5.1** Energy balance of a metallurgical complex

| Balance items | Mass | Energy [MWh] | | Exergy [MWh] | |
|---|---|---|---|---|---|
| | [t] | [$I_{ch}$] | [$I_q$] | [$E_{ch}$] | [$E_q$] |
| INPUT | | | | | |
| Coking coals | 3.471 | 22.317 | 0.0 | 21.499 | 0.0 |
| Coal concentrate | 0.034 | 0.405 | 0.0 | 0.368 | 0.0 |
| Coal at cogeneration plant | 0.064 | 0.368 | 0.0 | 0.357 | 0.0 |
| Coal for blast furnace | 0.013 | 0.094 | 0.0 | 0.090 | 0.0 |
| Fuel oil | 0.08 | 0.985 | 0.0 | 0.920 | 0.0 |
| Metal-bearing materials | 2.125 | 0.479 | 0.0 | 0.420 | 0.0 |
| Fluxes | 0.617 | 0.0 | 0.0 | 0.0 | 0.0 |
| Blast furnace additives | 0.110 | 0.221 | 0.0 | 0.200 | 0.0 |
| Steel ingots | 0.392 | 0.803 | 0.0 | 0.721 | 0.0 |
| Steel slabs | 0.190 | 0.387 | 0.0 | 0.350 | 0.0 |
| Tar | 0.028 | 0.317 | 0.0 | 0.304 | 0.0 |
| Reagents and other materials | | 0.051 | 0.0 | 0.051 | 0.0 |
| TOTAL: | – | 26.427 | 0.0 | 25.280 | 0.0 |
| OUTPUT | | | | | |
| Rolled stock | 1.0 | 2.044 | 0.0 | 1.847 | 0.0 |
| Pig iron, conversion of pig iron | 0.574 | 1.325 | 0.0 | 1.215 | 0.0 |
| to cast iron | 0.331 | 0.762 | 0.0 | 0.700 | 0.0 |
| Wastes of coal preparation | 0.112 | 0.542 | 0.0 | 0.527 | 0.0 |
| Coke | 0.493 | 3.716 | 0.0 | 3.694 | 0.0 |
| Pitch coke | 0.035 | 0.320 | 0.0 | 0.320 | 0.0 |
| Coke and coking by-products | 0.141 | 0.817 | 0.0 | 0.780 | 0.0 |
| Fertilizers | 0.024 | 0.0 | 0.0 | 0.0 | 0.0 |
| Granulated slag | 0.499 | 0.143 | – | 0.086 | – |
| Heat | | – | 0.560 | – | 0.078 |
| Electricity | | 0.009 | – | 0.009 | – |
| TOTAL: | | 9.678 | 0.560 | 9.178 | 0.078 |
| | | 10.238 | | 9.256 | |
| Energy and exergy losses | | 16.189 | | 16.024 | |

$\eta_{en} = 0.388$; $\eta_{ex} = 0.366$

with its own boundaries. This means that both energy flows supplied to the system from external sources and from other shops of the complex are also considered.

The tables present the material balances and balances of different energy forms. Electricity is shown in the columns with the chemical energy and exergy (to avoid further expansion of already large tables). The exergy values are given for each energy flow; the exergy flows are not balances. The difference between the total exergy in the input part of the balance and its total in the output part shows the exergy losses due to irreversibility of the processes.

Material or energy that is transferred from the shop under consideration to another shop of the complex or to external consumers were assumed to be useful.

**Coal-preparation plant.** Material and energy balances of the coal-preparation plant are given in Table 5.2. It shows that 3.47 tons of coking coals are fed into the coal preparation process which yields about 1.92 tons of concentrate for coking and 0.680 ton of intermediate product and slime. Coke-oven gas and an intermediate product are used as fuel to dry the concentrate, electricity is used for lighting and driving machines (pumps, thickeners, centrifuges) and heat energy is used for production needs. The values of energy and exergy efficiencies equal $\eta_{en} = 83.0\,\%$ and $\eta_{ex} = 82.7\,\%$, respectively.

**Coke ovens.** According to the material and energy balances of the coke ovens (Table 5.3), 1.953 tons of concentrate (and 0.034 ton of externally supplied concentrate) are used in the coking process which yields 1.469 tons of coke and 0.398 ton of uncleaned coke-oven gas. The coke batteries are heated by the coke-oven and blast-furnace gases. The energy and exergy efficiencies of the coke ovens are $\eta_{en} = 89.2\,\%$ and $\eta_{ex} = 91.2\,\%$.

**Chemical production.** A total of 0.411 ton of uncleaned coke-oven gas from the coke and pitch coke batteries and 0.028 ton of tar from external sources are fed into the chemical production process, yielding 0.257 ton of cleaned coke-oven gas, 0.019 ton of benzene, 0.011 ton of mid-temperature cinder, 0.038 ton of pitch coke, and other products (Table 5.4). Cleaning the coke-oven gas and trapping out its valuable components consume a total of about 0.019 ton of sulphuric acid, absorbent oil, soda ash and caustic soda and other reagents as well as electricity (0.025 MWh) and a large amount of heat (0.425 MWh). The energy and exergy efficiencies of the chemical production equal $\eta_{en} = 74.4\,\%$ and $\eta_{ex} = 82.8\,\%$, respectively.

**Agglomeration plant.** Table 5.5 presents the material and energy balances of the agglomeration plant which includes a shop for ore sintering and storehouses for the ore and concentrate.

About 1.96 tons of the iron ore, 0.058 ton of other iron-containing raw materials (iron chips, cinder, blast-furnace dust), recycled fine agglomeration particles as well as various fluxes (0.476) ton) and wast coke (0.132 ton) are used in the agglomeration process.

**Table 5.2** Energy balance of the coal preparation plant[a]

| Balance items | Mass | Energy [MWh] | | Exergy [MWh] | |
|---|---|---|---|---|---|
| | [t] | [$I_{ch}$] | [$I_q$] | [$E_{ch}$] | [$E_q$] |
| INPUT | | | | | |
| Coking coals | 3.47 | 22.317 | 0.0 | 21.499 | 0.0 |
| Coke-oven gas | 0.0016 | 0.021 | 0.0 | 0.018 | 0.0 |
| Coal, intermediate product | (0.098) | (0.56) | 0.0 | (0.542) | 0.0 |
| Heat | – | 0.0 | 0.031 | 0.0 | 0.011 |
| Electricity | – | 0.048 | 0.0 | 0.048 | 0.0 |
| Flotation reagents and auxiliary materials | ~ 0.01 | ~ 0.026 | 0.0 | ~ 0.024 | 0.0 |
| Compressed air | 0.004 | 0.0 | 0.0 | 0.0 | 0.0 |
| Blast air | ~ 0.692 | 0.0 | 0.0 | 0.0 | 0.0 |
| TOTAL: | 4.179 | 22.412 | 0.031 | 21.589 | 0.011 |
| | | 22.443 | | 21.6 | |
| OUTPUT | | | | | |
| *Coal burden for coking | 1.919 | 15.28 | 0.0 | 14.728 | 0.0 |
| *Intermediate product for external consumers | 0.662 | 3.196 | 0.0 | 3.066 | 0.0 |
| *Slime | 0.016 | 0.097 | 0.0 | 0.086 | 0.0 |
| Losses of coal and auxiliary materials | 0.859 | 3.212 | 0.0 | 2.933 | 0.0 |
| Heat lost with waste gases of driers | 0.723 | 0.0 | 0.442 | 0.0 | 0.0 |
| Heat losses of equipment and buildings | – | 0.0 | 0.216 | 0.0 | 0.0 |
| TOTAL: | 4.179 | 21.785 | 0.658 | 20.813 | 0.0 |
| | | 22.443 | | 20.813 | |

$\eta_{en} = 0.830$; $\eta_{ex} = 0.827$ .

[a]Here and in subsequent tables, values in parentheses are not included in the total; useful materials and products are marked by an asterisk

**Table 5.3** Energy balance of coke ovens

| Balance items | Mass | Energy [MWh] | | Exergy [MWh] | |
|---|---|---|---|---|---|
| | [t] | [$I_{ch}$] | [$I_q$] | [$E_{ch}$] | [$E_q$] |
| INPUT | | | | | |
| Burden for coking | 1.953 | 15.685 | 0.0 | 15.095 | 0.0 |
| incl. concentrate from external sources | (0.034) | (0.405) | 0.0 | (0.368) | 0.0 |
| Blast furnace gas (heating) | 0.475 | 0.355 | 0.0 | 0.320 | 0.0 |
| Coke-oven gas (heating) | 0.082 | 1.090 | 0.0 | 0.928 | 0.0 |
| Blast air | 1.687 | 0.0 | 0.0 | 0.0 | 0.0 |
| Water for gas cooling (recirculation) | (6.45) | 0.0 | 0.322 | 0.0 | 0.088 |
| TOTAL: | 4.197 | 17.130 | 0.322 | 16.343 | 0.088 |
| | | 17.452 | | 16.431 | |
| OUTPUT | | | | | |
| *Blast-furnace coke | 1.469 | 11.046 | (0.619) | 10.972 | (0.479) |
| *Coke-oven gas, uncleaned, cooled | 0.398 | 4.507 | 0.016 | 4.013 | 0.003 |
| Heat removed by steam-water mixture after coke-oven gas cooling | – | 0.0 | 0.538 | 0.0 | 0.150 |
| Heat losses with gases after recuperators | 2.288 | 0.0 | 0.184 | 0.0 | 0.092 |
| Heat losses during coke quenching | – | – | 0.619 | – | 0.479 |
| Heat losses from | – | – | 0.124 | – | 0.0 |
| Losses of coke-oven gases and burning losses | 0.042 | 0.418 | – | 0.356 | – |
| TOTAL: | 4.197 | 15.971 | 1.481 | 15.341 | 0.636 |
| | | 17.452 | | 15.977 | |

$\eta_{en} = 0.892$; $\eta_{ex} = 0.912$ .

**Table 5.4.** Energy balance of chemical production

| Balance items | Mass | Energy [MWh] | | Exergy [MWh] | |
|---|---|---|---|---|---|
| | [t] | [$I_{ch}$] | [$I_q$] | [$E_{ch}$] | [$E_q$] |
| INPUT | | | | | |
| Coke-oven gas from coke and coked-pitch batteries | 0.411 | 4.679 | 0.016 | 4.161 | 0.0 |
| Cleaned coke-oven gas for auxiliary needs | 0.006 | 0.087 | 0.0 | 0.075 | 0.0 |
| Blast air | 0.087 | 0.0 | 0.0 | 0.0 | 0.0 |
| Tar from outside | 0.028 | 0.317 | 0.0 | 0.304 | 0.0 |
| Circulating water from coke-batteries | (6.45) | 0.0 | 0.538 | 0.0 | 0.150 |
| Sulphuric acid, etc. | 0.017 | 0.005 | 0.0 | 0.005 | 0.0 |
| Absorption oil | 0.002 | 0.019 | 0.0 | 0.017 | 0.0 |
| Electricity | – | 0.025 | – | 0.025 | – |
| Heat | – | 0.0 | 0.425 | 0.0 | 0.118 |
| Compressed air | (0.012) | – | 0.0 | – | 0.005 |
| TOTAL: | 0.551 | 5.132 | 0.979 | 4.587 | 0.273 |
| | | 6.111 | | 4.860 | |
| OUTPUT | | | | | |
| *Coke-oven gas cleaning* | | | | | |
| *Cleaned coke-oven gas, including gas from coked-pitch batteries | 0.257 | 3.388 | 0.0 | 2.901 | 0.0 |
| Ammonium sulphate | 0.024 | 0.0 | 0.0 | 0.0 | 0.0 |
| Return cooled water | (6.45) | 0.0 | 0.322 | 0.0 | 0.088 |
| *Benzene retrieval* | | | | | |
| *Benzenes | 0.019 | 0.228 | 0.0 | 0.221 | 0.0 |
| *Toluene, xylene, solvent | 0.001 | 0.017 | 0.0 | 0.016 | 0.0 |
| Others | 0.001 | 0.011 | 0.0 | 0.011 | 0.0 |
| *Residues | 0.002 | 0.021 | 0.0 | 0.019 | 0.0 |
| *Tar distillation* | | | | | |
| *Absorption oil | 0.003 | 0.033 | 0.0 | 0.030 | 0.0 |
| *Average-temperature coked pitch | 0.11 | 0.105 | 0.0 | 0.105 | 0.0 |
| *Sleeper-impregnation oil | 0.015 | 0.168 | 0.0 | 0.152 | 0.0 |
| *Anthracene | 0.016 | 0.171 | 0.0 | 0.161 | 0.0 |
| *Pressed naphtalene | 0.005 | 0.059 | 0.0 | 0.058 | 0.0 |
| *Carbolic light oil | 0.001 | 0.009 | 0.0 | 0.009 | 0.0 |
| *Coked pitch shop* | | | | | |
| *Coked pitch | 0.038 | 0.337 | 0.0 | 0.339 | 0.0 |
| Coked pitch gas | (0.013) | (0.172) | – | (0.149) | – |
| Losses of gas and by-products | 0.045 | 0.048 | 0.0 | 0.390 | 0.0 |
| Heat loss in cooling towers, from equipment surfaces, with materials and waste gases | 0.113 | 0.0 | 0.794 | 0.0 | 0.0 |
| TOTAL: | 0.551 | 4.995 | 1.116 | 4.412 | 0.088 |
| | | 6.111 | | 4.500 | |

$\eta_{en} = 0.744; \quad \eta_{ex} = 0.828$ .

**Table 5.5** Energy balance of the agglomeration plant

| Balance items | Mass | Energy [MWh] | | Exergy [MWh] | |
|---|---|---|---|---|---|
| | [t] | $[I_{ch}]$ | $[I_q]$ | $[E_{ch}]$ | $[E_q]$ |
| INPUT | | | | | |
| Iron ore | 1.970 | 0.294 | 0.0 | 0.256 | 0.0 |
| Recycled fine-particle agglomerates | (0.950) | (0.142) | (0.0) | (0.124) | 0.0 |
| Other iron-containing materials | 0.058 | 0.088 | 0.0 | 0.084 | 0.0 |
| Fluxes, limestone, dolomite | 0.476 | 0.0 | 0.0 | 0.0 | 0.0 |
| Coke and coke breeze | 0.132 | 1.053 | 0.0 | 1.046 | 0.0 |
| Moisture of oven charge | 0.339 | 0.0 | 0.0 | 0.0 | 0.0 |
| Intermediate product | 0.004 | 0.023 | 0.0 | 0.022 | 0.0 |
| Blast furnace gas | 0.074 | 0.055 | 0.0 | 0.050 | 0.0 |
| Coke-oven gas | 0.007 | 0.097 | 0.0 | 0.083 | 0.0 |
| Air for sintering and cooling | 29.455 | 0.0 | 0.0 | 0.0 | 0.0 |
| Electricity | – | 0.087 | – | 0.087 | – |
| Heat | | – | 0.042 | – | 0.014 |
| Compressed air | (0.014) | – | 0.0 | – | 0.0 |
| TOTAL: | 32.515 | 1.698 | 0.042 | 1.625 | 0.014 |
| | | 1.740 | | 1.642 | |
| OUTPUT | | | | | |
| *Usable agglomerate, cooled | 2.092 | 0.357 | 0.0 | 0.292 | 0.0 |
| Recycled agglomerate and dust | (0.950) | (0.142) | (0.0) | (0.124) | (0.0) |
| Gas drawoff by exhausters | 16.450 | – | ~ 0.893 | – | 0.258 |
| Heat lost with air | 13.973 | – | 0.368 | – | 0.075 |
| Heat lost with cooling water | – | – | 0.048 | – | 0.0 |
| Heat losses from equipment surfaces | – | – | ~ 0.074 | – | 0.0 |
| TOTAL: | 32.515 | 0.357 | 1.383 | 0.292 | 0.331 |
| | | 1.740 | | 0.625 | |

$\eta_{en} = 0.205$; $\eta_{ex} = 0.178$ .

The process yields 2.092 tons of usable agglomerate. A total of 0.081 ton of the blast-furnace and coke-oven gases is consumed to ignite the oven charge in the hearth and 0.004 ton of the intermediate product in the ore sintering shop. About 0.087 MWh of electricity is consumed to drive the agglomeration machines flue gas extractors and exhausters, lighting and other needs and 0.042 MWh of heat energy is used for non-production needs.

The ratio of energy carriers consumed in the process is $I_r : I_f : E_{el} : I_q =$ 0.220 : 0.705 : 0.05 : 0.025, energy of coke being taken as the chemical energy of fuel. The energy and exergy efficiencies of the agglomeration process are not high, $\eta_{en} = 20.5\,\%$ and $\eta_{ex} = 17.8\,\%$.

**Blast-furnace shop.** The burden of the blast furnaces includes agglomerate passed on from the agglomeration shop as well as iron and manganese ores, iron pellets (0.109 ton), scrap, iron chips, welding slag (a total of 0.074 ton), ferrosilicate and other alloys (0.013 ton), fluxes (0.030 ton) and coke (after fine particles are removed, 0.76 ton). Coal dust and fuel oil are blasted into the furnaces (0.015 ton) to conserve coke (Table 5.6).

The process yields 1.265 tons of iron, 2.946 tons of blast-furnace gas, 0.498 ton of granulated slag, 0.024 ton of iron scrap.

Gas cleaners, casting machines and a slag processing plant are also included in the blast-furnace shop.

About 0.854 ton of the blast-furnace gas and 0.016 ton of the coke-oven gas are burnt in the Cowper hot blast stoves for heating the blast air. For lighting in the shops, driving casting machines, and for non-production needs of shops, 0.013 MWh of electricity and 0.082 MWh of heat are consumed.

The ratio of energy carriers consumed in the blast-furnace process is $I_r : I_f : E_{el} : I_q = 0.073 : 0.090 : 0.002 : 0.015$.

The energy and exergy efficiencies of the blast-furnace process are rather high, $\eta_{en} = 74.4\,\%$ and $\eta_{ex} = 69.3\,\%$.

**Open-hearth shop.** The open-hearth shop includes an ingot mould, scrap yard, and a room for dolomite roasting. Table 5.7 shows that open-hearth furnaces operate on liquid iron with a temperature of 1300°C (0.351 ton or 42 % of all the processed materials) and scrap (0.325 ton or 39 %). In addition, ore and agglomerate (0.053 ton), cinder (0.014 ton), lime, limestone, dolomite (0.374 ton), ferro-alloys and deoxidizers (0.009 ton) are loaded into the furnace. The process yields 0.648 ton of commercial steel, 0.051 ton of steel wastes and 0.204 MWh of heat from waste-heat boilers and evaporative-cooling systems of the open-hearth furnaces. Open-hearth and dolomite roasters are fueled by coke-oven gas (0.04 ton) and fuel oil (0.042 ton). Heat and electricity consumed for non-production purposes amount to 0.094 MWh and 0.012 MWh, respectively. The ratio of energy carriers consumed by the shop is

$$I_r : I_f : E_{el} : I_q = 0.53 : 0.39 : 0.004 : 0.076\,.$$

The energy and exergy efficiencies of the open-hearth process are $\eta_{en} = 61.8\,\%$ and $\eta_{ex} = 58.3\,\%$.

**Table 5.6.** Energy balance of the blast-furnace shop

| Balance items | Mass | Energy [MWh] | | Exergy [MWh] | |
|---|---|---|---|---|---|
| | [t] | $[I_{ch}]$ | $[I_q]$ | $[E_{ch}]$ | $[E_q]$ |
| INPUT | | | | | |
| Agglomerate | 2.092 | 0.357 | 0.0 | 0.292 | 0.0 |
| Ore | 0.109 | 0.019 | 0.0 | 0.016 | 0.0 |
| Recycled scrap | 0.007 | 0.014 | 0.0 | 0.013 | 0.0 |
| Welding slag | 0.007 | 0.015 | 0.0 | 0.014 | 0.0 |
| Iron chips and additives | 0.060 | 0.137 | 0.0 | 0.125 | 0.0 |
| Ferrosilicates, ferroalloys | 0.013 | 0.066 | 0.0 | 0.063 | 0.0 |
| Fluxes | 0.030 | 0.0 | 0.0 | 0.0 | 0.0 |
| Coke | 0.828 | 6.664 | 0.0 | 6.605 | 0.0 |
| Blast air | 2.607 | 0.0 | 0.044 | 0.0 | 0.088 |
| Coal dust (into blast-furnaces) | 0.013 | 0.094 | 0.0 | 0.90 | 0.0 |
| Fuel oil (into blast-furnaces) | 0.002 | 0.022 | 0.0 | 0.021 | 0.0 |
| Blast-furnace gas | 0.854 | 0.641 | 0.0 | 0.578 | 0.0 |
| Coke gas (into air heater) | 0.016 | 0.211 | 0.0 | 0.181 | 0.0 |
| Air for burning | 1.065 | 0.0 | 0.0 | 0.0 | 0.0 |
| Electricity | – | 0.013 | – | 0.013 | – |
| Heat | | – | 0.082 | – | 0.022 |
| Compressed air | (0.074) | 0.0 | 0.0 | 0.0 | 0.0 |
| TOTAL: | 7.703 | 8.253 | 0.126 | 8.011 | 0.110 |
| | | 8.379 | | 8.121 | |
| OUTPUT | | | | | |
| *Iron | 1.265 | 2.946 | (0.471) | 2.710 | (0.401) |
| including: | | | | | |
| conversion pig iron | (0.578) | | | | |
| cast pig iron | (0.331) | | | | |
| liquid conversion pig iron | (0.356) | (0.817) | 0.118 | (0.751) | 0.100 |
| *Blast-furnace gas | 3.157 | 2.380 | (0.420) | 2.140 | (0.122) |
| Losses of blast-furnace gas | 0.433 | 0.311 | 0.0 | 0.282 | 0.0 |
| Blast-furnace liquid slag | 0.609 | (0.175) | (0.272) | (0.105) | (0.232) |
| *Processes slag | (0.498) | 0.143 | 0.0 | 0.086 | 0.0 |
| *Flue dust (trapped) | 0.028 | 0.055 | 0.0 | 0.040 | 0.0 |
| *Iron scrap (non-standard) | 0.024 | 0.055 | 0.0 | 0.046 | 0.0 |
| *Coke size screening | 0.068 | 0.538 | 0.0 | 0.515 | 0.0 |
| Waste gases of air-heaters | 2.121 | 0.0 | 0.325 | 0.0 | 0.0 |
| Heat losses in gas cleaner | – | 0.0 | 0.420 | 0.0 | 0.122 |
| Heat losses during iron casting | – | 0.0 | 0.353 | 0.0 | 0.301 |
| Losses during slag processing | – | 0.032 | 0.272 | 0.019 | 0.232 |
| Heat losses from surfaces and from water cooling system | – | 0.0 | 0.431 | 0.0 | 0.151 |
| TOTAL: | 7.703 | 6.460 | 1.919 | 5.839 | 0.906 |
| | | 8.379 | | 6.744 | |

$\eta_{en} = 0.744$; $\eta_{ex} = 0.693$ .

**Table 5.7** Energy balance of an open-hearth shop

| Balance items | Mass | Energy [MWh] | | Exergy [MWh] | |
|---|---|---|---|---|---|
| | [t] | [$I_{ch}$] | [$I_q$] | [$E_{ch}$] | [$E_q$] |
| INPUT | | | | | |
| Liquid conversion iron | 0.351 | 0.806 | 0.118 | 0.736 | 0.096 |
| Iron (wastes) | 0.022 | 0.051 | 0.0 | 0.047 | 0.0 |
| Conversion pig iron | 0.004 | 0.011 | 0.0 | 0.009 | 0.0 |
| Steel (wastes) | 0.290 | 0.593 | – | 0.534 | – |
| Ferroalloys, deoxidizers | 0.009 | 0.004 | – | 0.003 | – |
| Ore, agglomerate, limestone and other additives | 0.157 | 0.010 | 0.0 | 0.009 | 0.0 |
| Coke-oven gas | 0.040 | 0.534 | 0.0 | 0.456 | 0.0 |
| Fuel oil | 0.042 | 0.547 | 0.0 | 0.502 | 0.0 |
| Air | 0.943 | – | 0.0 | – | 0.0 |
| Electricity | – | 0.012 | – | 0.012 | – |
| Heat | – | – | 0.094 | – | 0.026 |
| Compressed air | 0.144 | – | 0.0 | – | 0.006 |
| Air suction | 0.953 | – | 0.0 | – | 0.0 |
| TOTAL: | 2.955 | 2.568 | 0.211 | 2.308 | 0.128 |
| | | 2.779 | | 2.436 | |
| OUTPUT | | | | | |
| *Open-hearth steel in ingots | 0.648 | 1.326 | 0.153 | 1.193 | 0.149 |
| *Steel wastes, losses by burning | 0.051 | 0.034 | 0.0 | 0.020 | 0.0 |
| Open-hearth slag | 0.110 | 0.029 | 0.044 | 0.025 | 0.026 |
| *Steam of waste-heat boilers and evaporating cooling system | – | 0.0 | 0.204 | – | 0.058 |
| Heat losses during casting and ingot transportation | – | 0.0 | 0.158 | – | 0.129 |
| Losses with gases after waste-heat boilers | 2.146 | – | 0.279 | – | 0.099 |
| Heat losses from surfaces | – | – | 0.552 | – | – |
| TOTAL: | 2.955 | 1.389 | 1.390 | 1.238 | 0.461 |
| | | 2.779 | | 1.699 | |

$\eta_{en} = 0.618$; $\eta_{ex} = 0.583$ .

**Hot rolling shop.** The rolling shop includes a slabbing mill and a sheet mill, pit furnaces, continuous furnaces with waste-heat boilers, and a recirculating water supply system. Steel with an average temperature of 590°C (0.648 ton), and steel from other plants in the form of ingots (0.392 ton), as well as slabs (0.190 ton) are sent to the rolling shop. Heating of the pit furnaces and continuous furnaces consumes 0.633 ton of blast-furnace gas and 0.061 ton of coke-oven gas. For the driving mills, lighting, and other needs of the shop, 0.096 MWh of electricity is consumed. The shop heating and airconditioning, non-production needs require 0.169 MWh of heat. As a result, the shop produces steel sheets (0.753 ton), steel slabs and blooms (0.247 ton), heat energy from the waste-heat boilers and evaporative cooling systems of the continuous furnaces (0.145 MWh).

The energy use level in the rolling shop is characterized by energy and exergy efficiencies of $\eta_{en}$ = 62.8 % and $\eta_{ex}$ = 62.2 % (Table 5.8).

**Energy supply service.** The energy supply service includes cogeneration plants with steam-air units, oxygen and compressor plants, a heat and water supply. The energy supply service must supply the shops of the complex with electricity and heat, blast and compressed air, oxygen, industrial and drinking water. Table 5.9 presents the balance of energy consumed and produced by the energy service.

Most of the fuel in the energy supply service is consumed by the cogeneration plant with steam-air units for production of electricity and heat as well as blast air for the blast-furnace process. The table shows that the cogeneration plant with steam-air units itself consumes about 22 % of its total production of blast-furnace gas (0.655 ton). Note that the cogeneration plant is a "buffer-type" consumer of blast-furnace and coke-oven gases. That is, the excess of these gases that is not used in the primary production processes is fed back into it. This considerably increases the utilization factor of these gases at the complex and reduces losses.

The input part of the balance presents the total amount of heat produced by all the recovery plants of the complex (0.353 MWh) which accounts for 35 % of the heat consumption of the complex; its output part includes the total production of electricity, heat, blast and compressed air, and oxygen. The energy and exergy efficiencies of the energy supply service are equal to $\eta_{en}$ = 48.3 % and $\eta_{ex}$ = 27.6 %.

Table 5.10 is based on the energy balances of the main shops to show the influence of each stage on the overall efficiency of the complex. The main shops of the complex and the values of losses and useful energy (from the standpoint of the entire production process) are listed. It also presents the contribution of each stage in the total energy losses of the considered complex.

The main part of the total energy losses occurs in the coal preparation plant, the blast-furnaces and the energy supply service. But this does not mean that the efficiency of the energy use in these shops is low. The value of the total energy losses in a given shop does not reflect the actual efficiency of the energy use and will not reveal possibilities for improving a shop process.

**Table 5.8** Energy balance of a hot rolling shop

| Balance items | Mass | Energy [MWh] | | Exergy [MWh] | |
|---|---|---|---|---|---|
| | [t] | [$I_{ch}$] | [$I_q$] | [$E_{ch}$] | [$E_q$] |
| INPUT | | | | | |
| Open-hearth steel, domestic | 0.648 | 1.325 | 0.153 | 1.193 | 0.148 |
| Steel from external sources, including ingots, | 0.392 | 0.803 | 0.0 | 0.721 | 0.0 |
| slabs | 0.190 | 0.387 | 0.0 | 0.350 | 0.0 |
| Blast-furnace gas | 0.633 | 0.475 | 0.0 | 0.427 | 0.0 |
| Coke-oven gas | 0.061 | 0.810 | 0.0 | 0.700 | 0.0 |
| Blast air | 1.554 | – | 0.0 | – | 0.0 |
| Air suction | 5.891 | – | 0.0 | – | 0.0 |
| Electricity | – | 0.096 | – | 0.096 | – |
| Heat | | – | 0.169 | – | 0.046 |
| Compressed air | (0.035) | | – | – | – |
| TOTAL: | 9.369 | 3.896 | 0.322 | 3.487 | 0.194 |
| | | 4.218 | | 3.681 | |
| OUTPUT | | | | | |
| *Steel slabs and blooms | 0.247 | 0.504 | 0.0 | 0.455 | 0.0 |
| *Rolled stock | 0.753 | 1.539 | 0.0 | 1.384 | 0.0 |
| *Wastes | 0.236 | 0.453 | 0.0 | 0.409 | 0.0 |
| Heat losses during product cooling | | – | 0.464 | – | 0.0 |
| *Steam from waste-heat boilers | (124.421) | – | 0.145 | – | 0.041 |
| Losses with waste gases | 8.133 | 0.0 | 0.515 | 0.0 | 0.191 |
| Heat losses from surfaces | – | – | 0.598 | – | 0.0 |
| TOTAL: | 9.369 | 2.496 | 1.722 | 2.248 | 0.232 |
| | | 4.218 | | 2.480 | |

$\eta_{en} = 0.626; \quad \eta_{ex} = 0.622$ .

**Table 5.9** Energy balance of the energy supply

| Balance items | Mass | Energy [MWh] | | Exergy [MWh] | |
|---|---|---|---|---|---|
| | [t] | $[I_{ch}]$ | $[I_q]$ | $[E_{ch}]$ | $[E_q]$ |
| INPUT | | | | | |
| Fuel, including | | | | | |
| fuel oil | 0.008 | 0.097 | – | 0.094 | – |
| blast-furnace gas | 0.655 | 0.491 | – | 0.442 | – |
| coke-oven gas | 0.031 | 0.412 | – | 0.355 | – |
| coal | 0.064 | 0.368 | – | 0.357 | – |
| intermediate product | 0.449 | 2.577 | – | 2.489 | – |
| Electricity, including | | | | | |
| for auxiliary needs | – | 0.048 | – | 0.048 | – |
| for heat production | – | 0.044 | – | 0.044 | – |
| for air blasting | – | 0.019 | – | 0.019 | – |
| for compressing air | – | 0.033 | – | 0.033 | – |
| for oxygen production | – | 0.005 | – | 0.005 | – |
| for water supply | – | 0.073 | – | 0.073 | – |
| Heat of recovery installations | | – | 0.353 | – | 0.103 |
| TOTAL: | 1.207 | 4.167 | 0.353 | 3.959 | 0.103 |
| | | 4.520 | | 4.062 | |
| OUTPUT | | | | | |
| *Electricity | – | 0.556 | – | 0.556 | – |
| *Heat | – | – | 1.561 | – | 0.439 |
| *Blast air | 2.872 | – | 0.048 | – | 0.096 |
| *Compressed air | 0.398 | – | 0.0 | – | 0.020 |
| *Nondrinking water | 0.167 | – | 0.010 | – | 0.009 |
| Losses with waste gases and losses due to incomplete burning | – | – | 0.652 | – | 0.204 |
| Losses with cooling water, air and due to imbalance | 0.003 | – | 1.693 | – | 0.0 |
| Tonnage oxgygen | 0.003 | – | 0.0 | – | 0.0 |
| TOTAL: | | 0.556 | 3.964 | 0.556 | 0.768 |
| | | 4.520 | | 1.324 | |

$\eta_{en} = 0.483$; $\eta_{ex} = 0.276$ .

**Table 5.10** Distribution of useful energy and losses among shops of the metallurgical complex

| SHOPS | Useful energy produced | | Energy losses | | |
|---|---|---|---|---|---|
| | [MWh] | % of total consumption | [MWh] | % of total consumption | % of total losses |
| Coal preparation shop | 0.542 | 2.050 | 3.870 | 14.620 | 23.930 |
| Coke batteries | 3.716 | 14.100 | 1.561 | 5.900 | 9.630 |
| Chemical production | 1.137 | 4.310 | 1.242 | 4.690 | 7.660 |
| Agglomeration shop | – | – | 1.383 | 5.220 | 8.540 |
| Blast-furnace shop | 2.230 | 8.440 | 2.144 | 8.100 | 13.230 |
| Open-hearth shop | – | – | 1.062 | 4.010 | 6.570 |
| Rolling | 2.044 | 7.750 | 1.577 | 5.960 | 9.750 |
| Energy supply service | 0.569 | 2.150 | 2.345 | 8.900 | 14.470 |
| Auxiliary shops | – | – | 1.005 | 3.800 | 6.220 |
| TOTAL: | 10.238 | 38.800 | 16.189 | 61.200 | 100.000 |

### 5.2.1 Energy Use Efficiency

As was noted in Sect. 2.3, to estimate the maximum possible energy improvement of a shop or of the entire production complex one can compare the parameters of the processes with the parameters of their idealized analogs. In this section we make such a comparison for a steelmaking plant.

The idealized analog of steelmaking is assumed to consist of two stages: a) iron production out of pure magnetite; b) steelmaking out of the produced iron. This idealized production is described by the following reactions:

a) iron production

$$I_q + Fe_3O_4 + 4C = Fe_3C + CO_2 + 2CO\ ;$$

b) steelmaking

$$Fe_3C + O_2 = 3Fe + CO_2 + I_q\ .$$

The material and energy balances of the idealized steelmaking processes are given in Fig. 5.3. The data allow the determination of the energy and exergy efficiencies of each stage and the entire process:

iron production: $\eta_{en}^{id} = 1.0$; $\eta_{ex}^{id} = 1.0$;

steelmaking: $\eta_{en}^{id} = 0.773$; $\eta_{ex}^{id} = 0.728$;

overall process: $\eta_{en}^{id} = 0.826$; $\eta_{ex}^{id} = 0.796$.

The relative efficiency of a process is the generalized characteristic of their maximum possible improvement. The relative efficiencies of the blast-furnace and open-hearth processes of the considered complex are

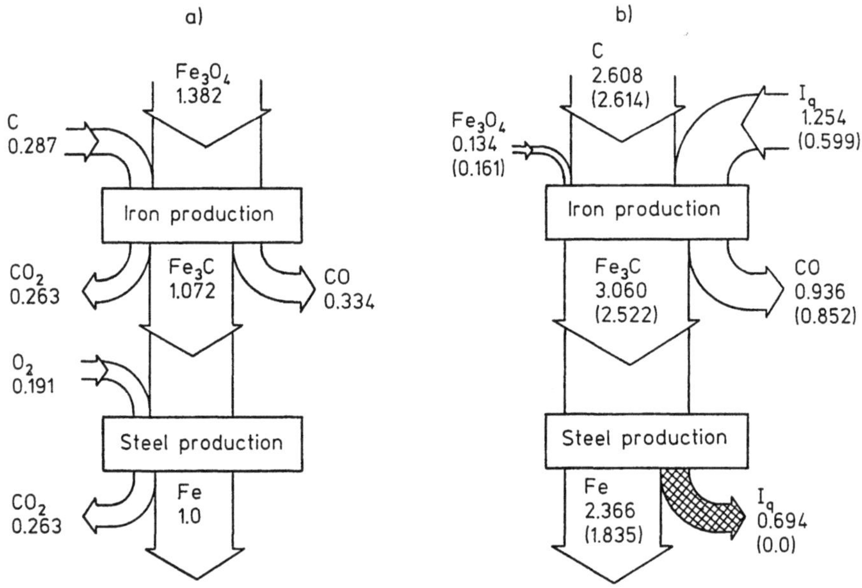


**Fig. 5.3.** Material balance [t] (*a*) and complete energy balance [kWh] (*b*) of the idealized analog of steelmaking (exergy values are given in parentheses)

blast-furnace process: $\eta_{en}^{rel} = 0.618/1.0 = 0.618$; $\eta_{ex}^{rel} = 0.583/1.0 = 0.583$;
open-hearth process: $\eta_{en}^{rel} = 0.744/0.773 = 0.962$; $\eta_{ex}^{rel} = 0.693/0.728 = 0.952$.

Apart from the maximum possible improvement of the individual processes and complexes on the whole, it is important to know the actual amount of energy saving reserves. To estimate this amount it is necessary to at least qualitatively analyze the energy losses revealed in the balances. This is the subject of the following section.

## 5.3 Energy Losses and Possible Secondary Energy Resources

The energy balance of shops in which the exergy values of all the items in its input and output allows a qualitative estimation of the kinds of losses encountered. The exergy values characterize the share of energy losses that can, theoretically, be utilized (the maximum value of secondary energy resources or theoretically applicable ones). The remaining part is the irretrievable losses.

Theoretically applicable secondary energy resources cannot be used completely in real conditions with currently available equipment. Therefore, in determining the actual recoverable losses, one should consider the best design of energy recycling equipment that can be applied in each particular case [5, 23, 24, 41, 42, 49, 73, 79, 94, 102, 118, 123, 124, 141, 192, 194, 196].

The results of such an analysis of the energy losses determined from energy balances of the main production stages of a metallurgical complex are given below.

**Coal preparation plant.** Losses of a coal preparation plant are mainly due to the losses of chemical energy with the burning part of tailings. This accounts for 21.0 % of the total losses of the complex. It is technically impossible to reduce these losses considerably. It is not advisable to use the heat of the waste gases from the driers, since this gas has a low temperature and high humidity. But the more advanced drying plants, for example, multiple-hearth fluidized bed furnaces, similar to those in use at alumina plants [21], can considerably reduce consumption of the primary fuel and, hence, the energy losses.

**Coke ovens.** The largest losses of chemical energy during coking are caused by leakages of the coke-oven gas and coke burning losses. This accounts for 3 % of the total losses at the complex. It is practically impossible to reduce these losses with available technology. Heat losses in coke-oven gas cooling are also high. Some of the gases can be used, but then the design of the coke-oven gas coolers must be drastically changed.

The mixture of the coke-oven and blast-furnace gases is at a temperature of 300 – 350°C after recuperators. The heat of these gases can be used as a low-potential secondary energy resources [79, 192]. However, the economic expediency of using this heat will be, to a great extent, determined by the number of external heat consumers in the region.

In our sample metallurgical complex, we use wet coke quenching, therefore, its heat is lost irretrievably. Installation of dry quenchers will allow this heat to be used for steam generation, as well as production of about 0.346 MWh of heat per ton of rolled steel, i.e., more than 50 % of the total heat removed with coke from the coke ovens. As much as 0.235 MWh of heat as steam at a pressure of 4.0 MPa and a temperature of 450°C can be produced per ton of commercial coke.

**Chemical production.** Cleaning the coke-oven gas and trapping out its valuable components have a low energy efficiency. Sublimation, separation, purification of raw material require repeated heating and cooling. Besides, these processes occur at a relatively low temperature (< 200°C). Usually the available production equipments are unable to use the heat released by these processes. Therefore, all the heat losses are practically fully unrecoverable.

Large flows and recirculation of the processed materials also cause considerable losses. The chemical energy of the lost gases and by-products is about 3 % of the total losses of the complex. It is difficult to reduce these values with the current technology.

**Agglomeration plant.** Agglomeration of the iron-bearing materials is far from being perfectly energy efficient. This shop is auxiliary, since the material it processes only acquires the mechanical properties required for further processing. The energy consumed in this process decreases the energy efficiency of the complex.

The largest fraction of energy losses in the agglomeration process is due to the waste gases. Utilization of this heat is technically difficult because of its low temperature and high dust and sulphurous anhydride content. Therefore, these and other losses of the agglomeration process are also said to be irretrievable.

Agglomeration can be replaced by the more energy efficient production of pellets using carbonate or cement binder [22]. This considerably reduces the energy consumption at this stage and, hence, in the complex as a whole.

**Blast-furnace process.** The main part of the energy losses at this stage are losses of the blast-furnace gases because of leaky loaders. They account for about 11.5 % of the overall output by the furnace or about 2.0 % of the total energy losses at the complex. Introduction of tight loaders can reduce the blast-furnace gas losses down to 3 – 5 % of their total output.

Energy losses during blast-furnace gas cleaning can be considerably reduced by partial use of the excess-pressure energy of furnace-top gases in the gas expanders. Heat of the waste gases from the Cowper hot-blast stoves can also be partially utilized [123].

Energy losses with cooling water can be fully utilized, if the furnaces are equipped with an evaporative-cooling system. Energy losses during transportation and processing of the blast-furnace slags are also considerable, but currently there are no appropriate ways for their utilization.

**Open-hearth process.** Although waste-heat boilers are typically installed behind open-hearth furnaces and dolomite roasters, the efficiency of using the waste gas heat is as low as 49 %. It can be increased to 60 % by reducing the air suction in the gas ducts.

Heat losses during ingot transportation from the open-hearth furnace to the rolling mills can be decreased with uninterrupted operation of rolling shops and the rail transport. The chemical energy of the open-hearth slags can be used during their processing with extraction of magnetic materials [17, 169].

**Rolling.** Heat of the waste gases from the pit and continuous furnaces is partially used in the economizers and waste-heat boilers. However, due to high air suction in the gas ducts (the air-fuel ratio before the waste-heat boiler of the continuous furnace is $\alpha = 2.6 - 3.0$), the utilization factor is 22 %. If the air suction is reduced, about 0.191 MWh of additional energy as steam can be produced per ton of steel, i.e., the heat utilization factor of the waste gases can become as high as 50 %.

Heat lost during the final product cooling can also be partially used, via heat exchangers [42, 73]. About 0.22 MWh of low-potential heat can be produced.

**Energy supply service and auxiliary shops.** Auxiliary shops supply the raw materials to the main processes, transport the products, remove wastes, manufacture dupplicate parts and refractories for repairing equipment, etc. Since the products of these shops are generally consumed within the complex, the energy consumed by them should be considered to be a part of the energy losses of the complex.

The same can be said about the energy supply service that provides the main shops of the complex with energy in different forms. But its products (hot water, steam, compressed air, etc.) are partly used by external consumers, therefore, the lost energy is less than that consumed by the amount of energy supplied to external consumers. It should be noted that energy losses of the energy supply service and auxiliary shops are mainly unrecoverable with the exception of heat lost with the waste gases of the cogeneration plant. These can be partially covered by equipment which is designed for low-potential heat.

The analysis of energy losses in the main shops of the metallurgical complex are given in Table 5.11.

The main energy losses in the different shops, the values of theoretically useful and irretrievable losses are listed. The last two columns show the chemical and heat energy losses that are technically (i.e., actually) usable. As can be seen, theoretically useful losses account for about 43.7 %, while usable losses amount to only about 12.8 % of the total energy losses of the complex. Hence, the energy efficiency of the complex could be increased up to 46.5 % if all the technically usable secondary energy resources were used.

## 5.4 Determination of the Economically Feasible Value of Using Thermal Secondary Energy Resources

Secondary energy resources are utilized at all the operating metallurgical plants, but not always efficiently. Nevertheless, it is not advisable to choose the optimal value of secondary energy resources utilization at existing plant by using the rather complex algorithm described in Chap. 4. Solution of the design problem is of greater interest.

Suppose that it is necessary to find the optimum extent of secondary energy resources utilization and the optimum capacities of the recovery installations after each SER source for a large, to be designed, metallurgical complex. This complex has a complete metallurgical cycle, and is similar in scheme and equipment to that of the example considered above. The capacity and curve for the secondary energy resources output, costs of recovery and furnace installations as well as techno-economic indices of power plants of the local power system are known for each secondary energy source.

The following aggregates are considered as sources of secondary energy resources at the design plant:

– three batteries of coke ovens with two dry quenchers with a maximum heat output of 50 and 35 Gcal/h (58.2 and about 40.7 MW) and constant annual curve of output;

**Table 5.11.** Analysis of energy losses in the main shops of a metallurgical complex [MWh per ton of steel]

| Losses | Yearly losses of energy and steel | | Exergy | | Anergy | | Theoretically recoverable | |
|---|---|---|---|---|---|---|---|---|
| | $I_{ch}$ | $I_q$ | $E_{ch}$ | $E_q$ | $B_{ch}$ | $B_q$ | Chemical | Heat |
| *Coal preparation plant* | | | | | | | | |
| Losses of coal and auxiliary materials | 3.212 | – | 2.933 | – | 0.279 | – | 0.0 | – |
| Losses with waste gases of driers | – | 0.442 | – | 0.228 | – | 0.214 | – | 0.171 |
| Heat losses from equipment surfaces | – | 0.216 | – | 0.0 | – | 0.216 | – | 0.0 |
| TOTAL: | 3.212 | 0.658 | 2.933 | 0.228 | 0.279 | 0.430 | 0.0 | 0.171 |
| | 3.870 | | 3.161 | | 0.709 | | 0.171 | |
| *Coke ovens* | | | | | | | | |
| Losses of coke-oven gas and by burning of coke | 0.418 | – | 0.356 | – | 0.062 | – | 0.132 | – |
| Heat losses during coke-oven gas cooling | – | 0.216 | – | 0.062 | – | 0.154 | – | 0.0 |
| Heat losses with waste gases | – | 0.184 | – | 0.092 | – | 0.092 | – | 0.069 |
| Heat loss during coke wet quenching | – | 0.619 | – | 0.479 | – | 0.140 | – | 0.346 |
| Heat losses from equipment surfaces | – | 0.124 | – | 0.0 | – | 0.124 | – | 0.0 |
| TOTAL: | 0.418 | 1.143 | 0.356 | 0.633 | 0.062 | 0.510 | 0.132 | 0.415 |
| | 1.561 | | 0.989 | | 0.572 | | 0.547 | |
| *Chemical production* | | | | | | | | |
| Losses of coke-oven gas and by-products | 0.448 | – | 0.390 | – | 0.058 | – | 0.145 | – |
| Heat losses from equipment surfaces | – | 0.794 | – | 0.0 | – | 0.794 | – | 0.0 |
| TOTAL: | 0.448 | 1.794 | 0.390 | 0.0 | 0.058 | 0.794 | 0.145 | 0.0 |
| | 1.242 | | 0.390 | | 0.852 | | 0.145 | |
| *Agglomeration plant* | | | | | | | | |
| Heat losses with waste gases | – | 0.893 | – | 0.253 | – | 0.635 | – | 0.0 |
| Heat losses with air cooling agglomerate | – | 0.368 | – | 0.075 | – | 0.293 | – | 0.0 |
| Heat losses from surfaces and during transportation | – | 0.122 | – | 0.0 | – | 0.122 | – | 0.0 |
| TOTAL: | – | 1.383 | – | 0.333 | – | 1.050 | – | 0.0 |
| | 1.383 | | 0.333 | | 1.050 | | | |

| | | | | | | | | |
|---|---|---|---|---|---|---|---|---|
| *Blast furnace process* | | | | | | | | |
| Losses of blast-furnace gas | 0.311 | – | 0.282 | – | 0.029 | – | 0.188 | – |
| Energy losses during slag processing | 0.032 | 0.272 | 0.019 | 0.232 | 0.013 | 0.040 | 0.0 | 0.0 |
| Heat energy losses during blast furnace gas cleaning | – | 0.420 | – | 0.122 | – | 0.298 | – | 0.059 |
| Heat losses with waste gases from Cowper stoves | – | 0.165 | – | 0.077 | – | 0.880 | – | 0.029 |
| Heat losses during iron casting and transportation | – | 0.353 | – | 0.301 | – | 0.052 | – | 0.0 |
| Heat losses from equipment surfaces and during transportation | – | 0.436 | – | 0.0 | – | 0.436 | – | 0.0 |
| Heat losses with water cooling of blast-furnaces | – | 0.155 | – | 0.155 | – | 0.0 | – | 0.155 |
| TOTAL: | 0.343 | 1.801 | 0.301 | 0.887 | 0.042 | 0.914 | 0.188 | 0.243 |
| | 2.144 | | 1.188 | | 0.956 | | 0.431 | |
| *Open-hearth process* | | | | | | | | |
| Losses with slag | 0.029 | 0.044 | 0.025 | 0.026 | 0.004 | 0.018 | 0.029 | 0.0 |
| Heat losses during casting and transportation of ingots | – | 0.158 | – | 0.129 | – | 0.029 | – | 0.079 |
| Heat losses with waste gases | – | 0.279 | – | 0.099 | – | 0.180 | – | 0.009 |
| Heat losses from equipment surfaces | – | 0.552 | – | 0.0 | – | 0.552 | – | 0.0 |
| TOTAL: | 0.029 | 1.033 | 0.025 | 0.254 | 0.004 | 0.779 | 0.029 | 0.178 |
| | 1.062 | | 0.279 | | 0.783 | | 0.207 | |
| *Rolling* | | | | | | | | |
| Heat losses during product cooling | – | 0.464 | – | 0.339 | – | 0.125 | – | 0.221 |
| Heat losses with waste gases | – | 0.515 | – | 0.191 | – | 0.324 | – | 0.191 |
| Heat losses from equipment surfaces | – | 0.598 | – | 0.0 | – | 0.598 | – | 0.0 |
| TOTAL: | – | 1.577 | – | 0.530 | – | 1.047 | – | 0.412 |
| | 1.577 | | 0.530 | | 1.047 | | 0.412 | |
| *Energy service* | | | | | | | | |
| Losses with waste gases | – | 0.652 | – | 0.204 | – | 0.448 | – | 0.153 |
| Losses with cooling water, from equipment surfaces and in heat networks | – | 1.693 | – | 0.0 | – | 1.693 | – | 0.0 |
| TOTAL: | – | 2.345 | – | 0.204 | – | 2.141 | – | 0.153 |
| | 2.345 | | 0.204 | | 2.141 | | 0.153 | |
| Total losses of the complex with account of auxiliary shops | 16.189 | | 7.074 | | 9.115 | | | |
| The same in % | 100.0 | | 43.7 | | 56.3 | | 12.8 | |

– two open-hearth furnaces with a capacity of 600 tons each and maximum capacity of heat flow in the waste gases of 15 Gcal/h (17.4 MW) each; the number of operating hours is 7350 a year;

– rotary furnace for dolomite roasting with 6000 operating hours a year and maximum capacity of heat flow in the waste gases of 10 Gcal/h (11.6 MW);

– four continuous furnaces with a maximum capacity of heat flow in the waste gases of 40 Gcal/h (46.5 MW) each; and 8760 operating hours of three furnaces with one standby furnace;

– pit furnaces with a maximum capacity of heat flow in the waste gases of 10 Gcal/h (11.6 MW) with 7000 operating hours.

Sources of secondary energy resources have a highly diverse and irregular output curve. Typical secondary energy resources output from an open-hearth furnace, a converter, and dry quenchers are given in Figs. 5.4 – 5.6. These outputs of secondary energy resources have been averaged over the periods of a technological cycle. Duration curves of the secondary energy resources output are obtained from statistical analysis of data of similar aggregates at operating plants and are shown in Fig. 5.7.

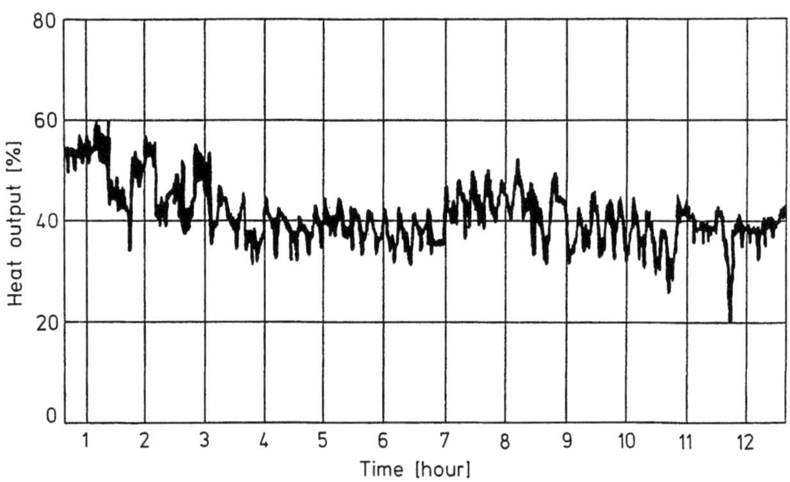


**Fig. 5.4.** Secondary energy output of the waste gases of an open-hearth furnace

According to our technique for choosing the economically expedient extent of secondary energy resource utilization, it is necessary to first obtain the equivalent characteristic of the SER utilization costs, i.e., to determine the dependence $Z_{SER} = f(L_{SER})$. Each value of $L_{SER}$ is assigned to an optimized group of heat recovery installations which involves the minimum cost of using this amount of secondary energy resources.

As noted earlier, any metallurgical complex usually has several sources of secondary energy resources whose combination has already been optimized. In

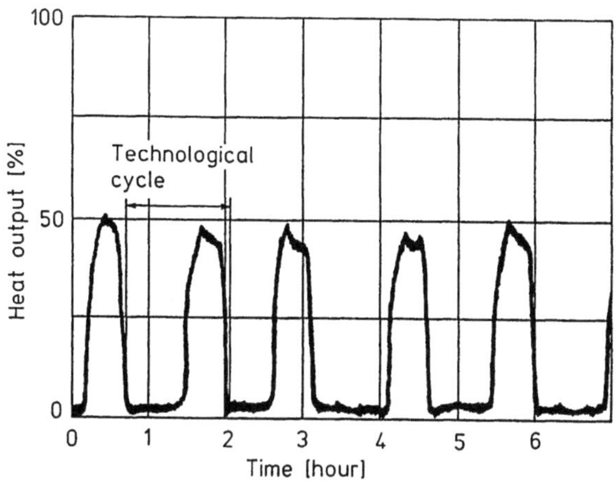


**Fig. 5.5.** Thermal secondary energy output of an oxygen converter

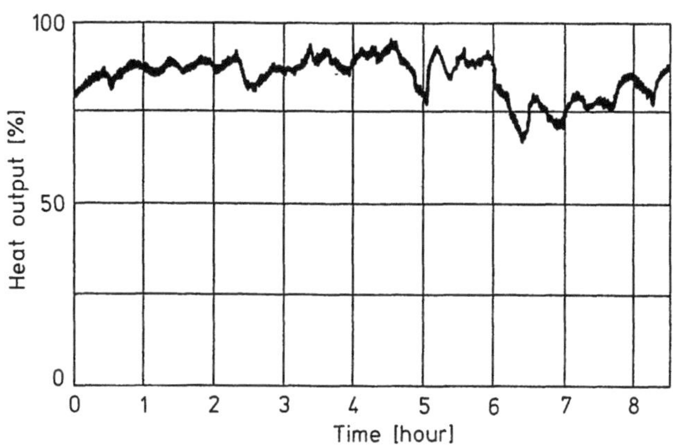


**Fig. 5.6.** Secondary energy output of a coke dry-quencher

other words, it is not necessary to determine the optimum capacity of such sources since it can be taken as the maximum possible one. These sources of secondary energy resources are not included in the calculation of $\underset{\min}{Z}_{\mathrm{SER}} = f(L_{\mathrm{SER}})$.

Such sources might include systems for evaporative cooling of structural components of the blast, open-hearth and continuous furnaces. The maximum invariable capacity of these sources in 100 Gcal/h (116.4 MW).

The optimum total capacity of recovery installations which, together with the heat and electricity supply systems, participate in meeting the heat loads of the region, were calculated to optimize the structure of energy supply system by using the algorithm described in Chap. 4. Indices characterizing the energy installations of the system such as the necessary capital investment into condensation and cogeneration power plants, industrial and peak boilers, their specific fuel consumption, etc., as well as preliminarily optimized costs of utilization of secondary

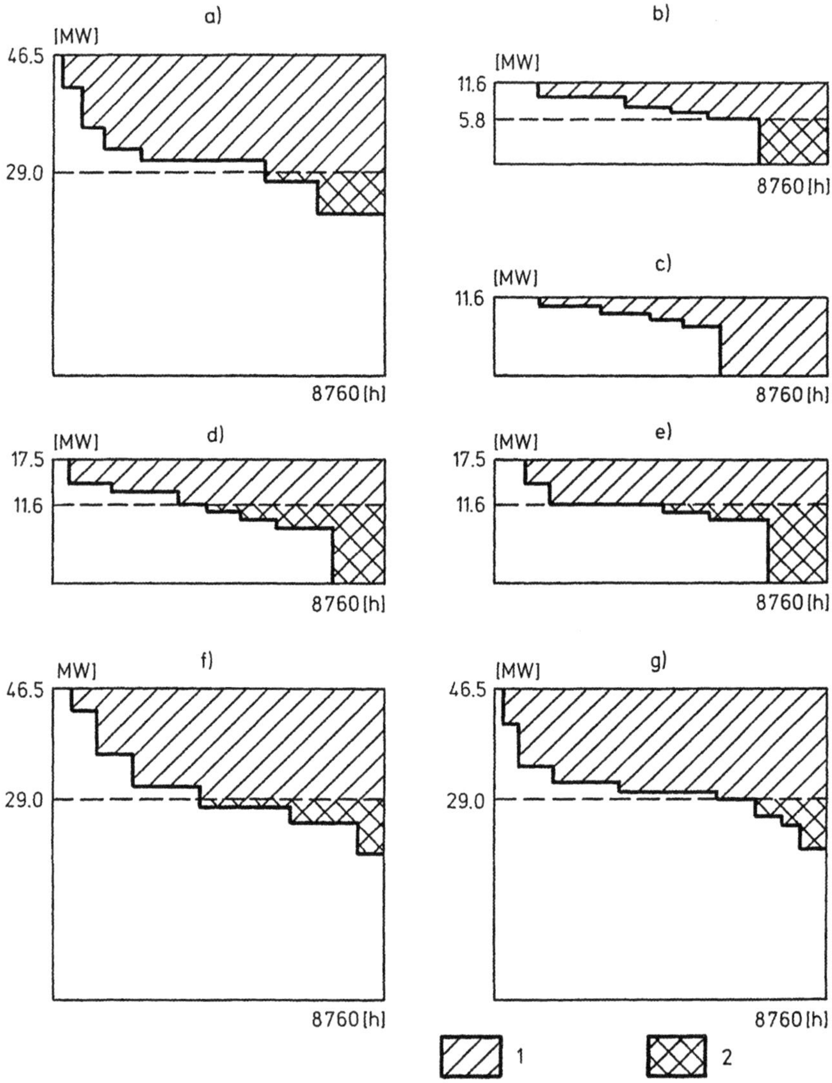


**Fig. 5.7.** Heat output of SER sources. 1 – heat production by furnace installations at the maximum value of SER utilization; 2 – same at the optimum capacity of recovery installations: *a*, *f*, *g* – continuous furnaces; *d*, *e* – open-hearth furnaces; *b* – heating furnaces; *c* – furnace for dolomite roasting

energy resources $\underset{\min}{Z}_{\text{SER}} = f(L_{\text{SER}})$ served as input data. The calculations also allowed the determination of the absolute value of economic effect due to using secondary energy resources at the complex. To do this the structure of the heat and electricity supply system was optimized for the variants with recovery installations in the energy balance and without them. The difference in the costs for these variants gives the value of the economic effect due to utilization of secondary energy resources at the complex, i.e., EF = $\Sigma Z'_{compl} - \Sigma Z''_{compl}$, where $\Sigma Z'_{compl}$ are the total minimized costs of the electricity and heat supply system without recovery installations; $\Sigma Z''_{compl}$ is the same with recovery units.

The cost of fuel for heat production substantially affects the optimum degree of secondary energy resources utilization. Calculations were performed for different the lowest and highest costs of fuel burnt in energy installations (from 15 to 45 rubles/t c.e.) and in furnace installations (from 15 to 70 rubles/t c.e.). Figure 5.8 shows the optimum degree of secondary energy resources utilization at a metallurgical complex and the achieved economic effect.

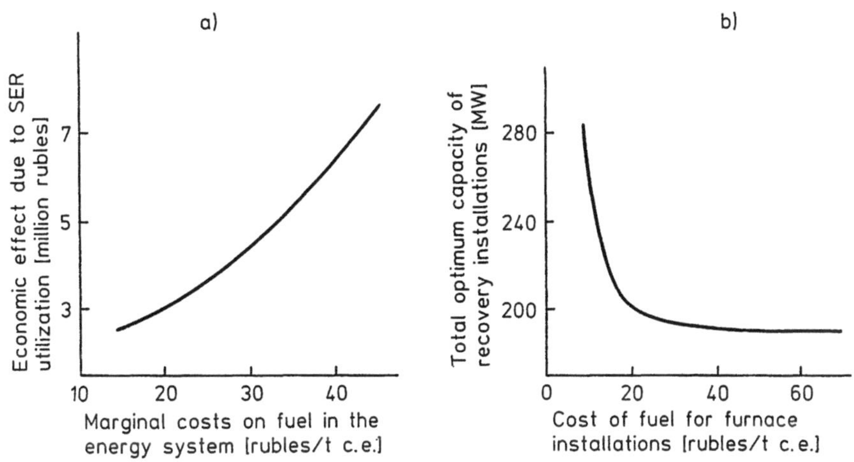


**Fig. 5.8.** Economic effect due to secondary energy resources utilization versus the fuel costs in an energy supply system (*a*) and optimum utilized capacity versus the fuel cost in furnace installations (*b*)

For example, the optimum value of the total heat output of recovery installations at a sample metallurgical complex is calculated as 185 Gcal/h (215.2 MW). The minimum costs of this heat output correspond to the following combination of sources of secondary energy resources (Fig. 5.7):

– optimum capacity of two recovery installations using hot coke heat in gas expanders, 50 Gcal/h (58.2 MW) and 35 Gcal/h (40.7 MW), respectively;

– optimum capacity of waste-heat boilers of the waste gases from continuous furnaces, 25 Gcal/h (29.1 MW);

– optimum capacity of the waste-heat boilers after open-hearth furnaces, 10 Gcal/h (11.6 MW);

– capacity of the waste-heat boilers after pit furnaces is taken to be equal to 5 Gcal/h (5.8 MW);

– waste-heat boilers after the dolomite roasting furnace are not included.

# 6. Energy Use and Energy Efficiency Increase in Non-Ferreous Metallurgy

The special feature of non-ferrous metallurgy is that the chemical energy of the raw materials constitutes the major (and sometimes even greatest) part of the energy supply. It was the study of the energy efficiency of the processes in this industry that led to improvements in the techniques for drawing up and analyzing energy balances. Advantages of the methodology developed by the author are most vividly revealed in studies on the processes of non-ferrous metallurgy, particularly:

– drawing up the complete energy balance that takes into account the chemical energy of raw material and other non-combustibles, the exergy value is indicated for each energy flow;

– use of the ideal and idealized analogs of actual processes for determining their relative energy efficiency;

– application of the exergy criterion when sharing the energy consumption for realization of the complex processes.

## 6.1 Copper Production

Copper is produced from ores which can be sulphides, oxides or mixed species. Sulphide ores make up 80 % of all copper ores. Their main component is copper pyrite or chalcopyrite ($CuFeS_2$). Other sulphides of copper and iron (bornite – $Cu_3FeS_3$, covellite – CuS, chalcosite – $Cu_2S$, pyrite – $FeS_2$, etc.), zinc blende (ZnS), lead sulphide (PbS), noble metals, compounds of nickel, cobalt, arsenic, antimony, selenium and tellurium make up the rest of the ore. The copper content in preliminarily processed ores usually varies from 1 to 5 %. Using current technology it is unprofitable to process ores with a copper content of $< 0.3$ %.

Ores are concentrated by the method of selective flotation which allows the isolation of relatively pure monometallic concentrates. Copper concentrates usually contain from 8 to 35 % Cu, as well as zinc, lead and other elements. There are two methods of copper production: hydrometallurgical and pyrometallurgical. The choice of the method depends on the chemical and mineralogical composition of the ore, its copper content, and the fuel and electricity costs, etc. arising in the processing.

Pyrometallurgical methods are currently the most widely used. Oxide-bearing or preliminarily roasted ores are usually processed by the hydrometallurgical method. Hydrometallurgy based on acid and bacteria leaching of copper ores with subsequent electrolysis of the obtained liquors, should become important in the near future.

### 6.1.1 Production Scheme and Energy Balances in Reverberatory Smelting

This method of smelting is rather widely used despite some disadvantages, first of all unprofitability of sulphur recovery from the waste gases due to low $SO_2$ content.

The basic scheme of copper production using reverberatory smelting of concentrates is given in Fig. 6.1. It presents the main stages of the entire production process, from ore mining to electrolytic copper production.

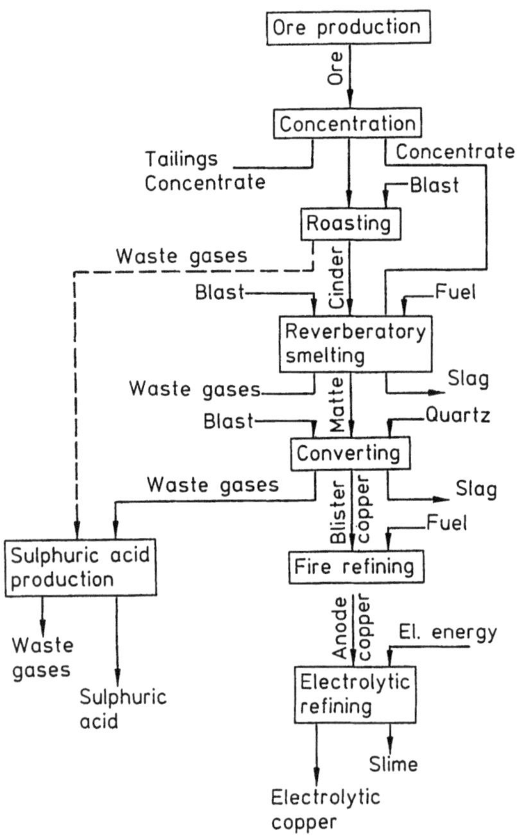


**Fig. 6.1.** Copper production using reverberatory smelting

**Concentration.** Ores with a copper content of less than 2% are concentrated. Concentration includes the following operations:

a) crushing and grinding;

b) concentration of the ground ore by flotation;
c) three-stage dehydrating: thickening, filtrating and drying.

**Concentrate roasting.** The sulphide ore is roasted to reduced the excess sulphur to an amount at which smelting will yield matte which can be efficiently converted to the end products.

Concentrates are usually roasted in fluidized bed furnaces. The roast gas produced contains $SO_2$ and is used for sulphuric acid production. The cinder produced is smelted to matte in the reverberatory furnaces.

**Reverberatory smelting.** Reverberatory smelting of copper concentrates can be done in two ways: smelting of concentrates after preliminary roasting or smelting of the raw dehydrated concentrates. Both have some advantages and disadvantages. Therefore, the choice of the method should be made in each particular case. Usually preference is given to the smelting of an unroasted concentrate.

A raw concentrate with a 10 % moisture content after dehydration is loaded into the reverberatory furnace in which it is smelted to copper matte with the addition of fluxes, liquid converter slag and recycled materials. The furnace is fueled with oil and coal dust to a temperature of 1500°C. In wet charge smelting, a small amount of iron is slagged; it is mainly transformed to matte. The transformation of copper and noble metals into matte in reverberatory smelting can be as high as 96 – 98.5 %.

**Converting.** The matte obtained in the reverberatory furnace must be further processed before copper can be extracted. The smelted copper matte is poured into a converter and blasted with air.

The blasting can be divided into two periods. During the first period, ferric sulphide and the major part of the impurities are oxidized. Almost pure copper sulphide containing 78 – 80 % Cu (white matter) is obtained. This is converted into copper during the second period.

The converting process yields blister copper containing 96 – 98 % Cu, converter slags and gases. Converter slags containing 1 – 5 % copper are poured back into reverberatory furnaces. The gases contain about 10 – 18 % $SO_2$ and are therefore sent on to sulphuric acid production.

**Converter copper refining.** Blister copper contains impurities such as Fe, Ni, S, Zn, As, Sb, Bi, Sn, Pb, Se, Te, Au, Ag. The presence of even low quantities of impurities in copper noticebly changes its quality; therefore, further refining, pyro- and electrolytics, is required.

Refined cathode copper contains 99.9 – 99.95 % Cu. This copper is partially used for industrial purposes but its major part is remelted in flame-contact or electric furnaces and cast into ingots or rods.

**Sulphuric acid production.** Copper smelteries produce sulphuric acid from the waste gases of the main copper production stages of concentrate roasting (if the method of smelting of roasted concentrates is used) and matte converting.

The contact method of sulphuric acid production includes three stages: (1) cleaning the gases from dust and impurities that adversely affect the catalysts; (2) contact oxidation of sulphur dioxide into sulphur trioxide ($2SO_2 + O_2 \rightarrow 2SO_3$); (3) absorption of sulphur trioxide by sulphuric acid.

This reaction in step (2) accelerated by catalysts. The most effective catalyst is the vanadium contact mass, containing, on the average, about 7 % $V_2O_5$. The optimum temperature of the process with this catalyst lies in the range of 400 – 600°C.

**The Energy Balance.** Consider, for example, a metallurgical complex operating according to the above production scheme, with reverberatory smelting of unroasted concentrates. The complex comprises mines, a concentration plant, a filtering-drying plant, reverberatory furnaces with a coal-dust processing plant, a converter, sulphuric acid production, pyrolytic and electrolytic refining. Apart from its own ores the complex processes concentrates, pellets and blister copper from other factories, as well as fluxes (limestone, quartz, etc.). Coal dust is used as an energy carrier at the reverberatory stage, fuel oil is used in the pyrolytic refining, and electric power is needed in the electrolytic refining. Heat for both industrial processes and public utility services is produced in the waste-heat boilers of the complex and a cogeneration plant located in the near vicinity. The complex produces cathode and anode copper, copper vitriol and sulphuric acid, obtained from the waste gases of the converting stage. All these products can be expressed in energy terms, as described in previous chapters, to allow an energy balance of the copper production process to be found (Table 6.1).

The table presents the amounts of input materials and fuel and the corresponding values of chemical energy and exergy per ton of electrolytic copper. Similar data are obtained for the other products of the complex. The input and output parts of the balance include other conventional energy forms: heat and electricity from other, external sources and those produced by the complex. The obtained data allow the determination of the energy use efficiency, characterized by energy and exergy efficiencies whose values are also given in Table 6.1.

One very important observation can be made: The considered copper production process, based on conventional energy carriers yields only 55.5 % of the total required energy, the remaining 44.5 % are contributed by the raw material. Thus, chemical energy of the raw material, which is usually neglected in analyses of energy efficiency, plays a significant role in copper production and in other energy-intensive production processes. Its neglect can result in gross errors.

For example, the total energy consumption for this well-established method of copper production is relatively stable. In long- or short-term forecasting it can be assumed to remain constant. Thus, the chemical energy of the raw material must be accounted for since its value is not a constant and depends on the composition of the raw material.

**Table 6.1** Energy balance of copper production (actual)

| Balance items | Mass [t] | Energy [MWh] | Exergy [MWh] |
|---|---|---|---|
| INPUT | | | |
| Sulphide ore | 41.5 | 4.64 | 4.16 |
| Concentrates from external suppliers | 1.79 | 2.07 | 1.74 |
| Pellets | 0.41 | 0.445 | 0.375 |
| Blister copper from external suppliers | 0.255 | 0.247 | 0.195 |
| Limestone | 0.49 | 0.0 | 0.0 |
| Quartz and other ores | 0.29 | 0.05 | 0.04 |
| Coal[a], t c.e. | 0.795 | 6.5 | 6.3 |
| Fuel oil, t c.e. | 0.11 | 0.89 | 0.83 |
| Electricity | – | 1.73 | 1.73 |
| Heat energy from external suppliers | – | 0.14 | 0.02 |
| TOTAL | 45.64 | 16.712 | 15.39 |
| OUTPUT | | | |
| Cathode copper | 1.0 | 0.981 | 0.763 |
| Commercial anodes | 0.098 | 0.096 | 0.075 |
| Copper vitriol | 0.028 | 0.004 | 0.002 |
| Sulphuric acid (monohydrate) | 0.76 | 0.124 | 0.084 |
| Heat of waste-heat boilers for external consumers | – | – | – |
| Useful energy, total | | 1.205 | 0.924 |
| Energy and exergy losses | | 15.507 | 14.466 |

$\eta_{en} = 1.205/16.712 = 0.072$; $\eta_{ex} = 0.924/15.39 = 0.06$

[a] tons of coal equivalent

The energy balance shows that the existing level of energy use in copper production based on reverberatory smelting yields $\eta_{en} = 7.2\,\%$ and $\eta_{ex} = 6\,\%$ with the following ratio of energy carriers:

$$I_r : I_f : E_{el} : I_q = 0.445 : 0.443 : 0.104 : 0.008\,.$$

To evaluate the degree of perfection of this real production process, its energy and exergy efficiencies were compared to those of an idealized analog. This analog was taken as the process of copper production from pure chalcopyrite ($CuFeS_2$):

$$2CuFeS_2 + 5O_2 + SiO_2 = 2Cu + Fe_2SiO_4 + 4SO_2 + \Delta H_r\,.$$

Because the complex includes sulphuric acid production, this process should also be taken into consideration in the idealized analog. This process is described by the following reaction:

$$2SO_2 + O_2 + 2H_2O = 2H_2SO_4 + \Delta H_r \,.$$

Figure 6.2 presents diagrams of material and energy balances of an idealized copper production process. If sulphur is not recovered and copper is the only useful product, then the energy and exergy efficiencies are $\eta_{en}^{id} = (981/7760) \times 100 = 12.6\,\%$; $\eta_{ex}^{id} = (763/6000) \times 100 = 12.7\,\%$. If copper and sulphuric acid are produced, the energy and exergy efficiencies of the idealized analog are equal to $\eta_{en}^{id} = 23.8\,\%$ and $\eta_{ex}^{id} = 23.4\,\%$, respectively. The relative efficiencies of a real copper-production complex are thus $\eta_{en}^{rel} = 30.2\,\%$ and $\eta_{ex}^{rel} = 25.6\,\%$.

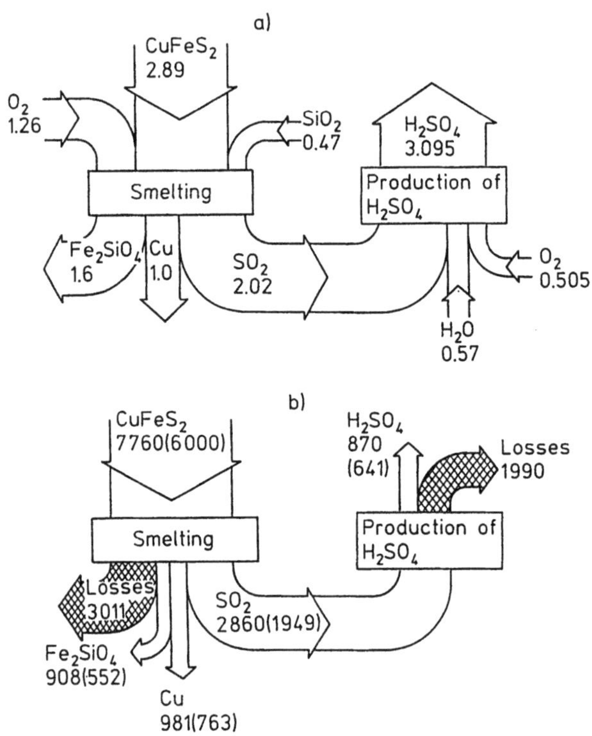


**Fig. 6.2.** Diagrams of material balance [t] (*a*) and complete energy balance [kWh] (*b*) of the idealized analog of copper production using reverberatory smelting (exergy values are given in parentheses)

In order to reveal the ways of increasing the energy efficiency of a complex, to identify the losses and their causes, and to evaluate the possiblity of using them as secondary energy resources, the energy balances of the individual main processes were drawn. Table 6.2 shows the results.

Losses of chemical and heat energy with the waste gases from reverberatory and anode furnaces and converters are of greatest interest to us in terms of the

**Table 6.2** Irretrievable losses in copper production of heat and chemicals

| Stages and types of losses | Actual losses [MWh] | | Calculated losses [MWh] | |
|---|---|---|---|---|
| | Chemical $I_{ch}/E_{ch}$ | Heat $I_q/E_q$ | Chemical $I_{ch}/E_{ch}$ | Heat $I_q/E_q$ |
| *Concentration plant* | | | | |
| Tailings | 1.725/1.55 | 0.0 | 1.725/1.55 | 0.0 |
| Losses from the equipment surfaces | 0.0 | 0.96/0.0 | 0.0 | 0.96/0.0 |
| *Filtering-drying shop* | | | | |
| Losses with dusty, wet gases and from surfaces | 0.0 | 0.48/0.0 | 0.0 | 0.48/0.0 |
| *Reverberatory shop with coal dust processing plant, oxygen and blast plants and waste-heat boilers* | | | | |
| Dump slag | 0.61/0.31 | 1.19/0.0 | 0.61/0.31 | 1.18/0.0 |
| Losses with waste gases | 0.734/0.61 | 3.528/1.9 | 0.0 | 0.0 |
| Losses from surfaces | – | 1.0/0.0 | – | 1.29/0.0 |
| *Production of liquid $SO_2$* | | | | |
| Losses with waste gases | – | – | 0.20/0.0 | 1.20/0.0 |
| Losses from surfaces | – | – | – | 1.64/0.0 |
| *Converter shop with waste-heat boilers and sulphuric acid production* | | | | |
| Losses with waste gases | 0.98/0.81 | 1.13/0.6 | 0.227/0.14 | 0.44/0.0 |
| Losses from surfaces | – | 1.0/0.0 | – | 1.61/0.0 |
| *Pyrolytic refining* | | | | |
| Losses with waste gases | – | 0.46/0.30 | – | 0.117/0.0 |
| Losses from surfaces | 0.12/0.0 | 0.52/0.0 | 0.12/0.0 | 0.53/0.0 |
| *Electrolytic refining* | | | | |
| Losses from surfaces | – | 1.08/0.0 | – | 1.08/0.0 |
| Energy losses, total: | 4.169 | 11.338 | 2.882 | 10.527 |
| | 15.507 | | 13.409 | |

multi-product processing of raw material, increasing of energy efficiency and environmental protection from harmful emissions. The values of losses given in the column labeled "actual" losses are derived for an actual production process which partially recovers the heat of the waste gases from reverberatory furnaces and removes an amount of $SO_2$ from the converter gases in the course of sulphuric acid production.

We calculated the energy balance of a copper production complex, whose production scheme, volume of production and volume of raw material used are similar to those of the plant conisdered in the preceding sections, but with more complete utilization of secondary energy resources and increased efficiency via oxygen-enriched air blast. It was taken into account that the use of thermal and chemical secondary energy resources is interrelated. For example, equipping produciton units with waste-heat boilers leads to a reduction in the volume of waste gases which in turn, increases their $SO_2$ concentration and hence, the efficiency of sulphur dioxide trapping.

The energy balance also took into acocunt the additional energy consumption from new processes utilizing thermal and chemical secondary energy resources. The results of these calculations are given in Table 6.3.

Note that the energy input in Tables 6.1 and 6.3 (last 4 items under "Input") are quite different. Energy supplied by coal is decreased due to better heating of the blast air in the air heaters that are installed after the waste-heat boilers. The increase in the oil and electricity consumption in Table 6.3 is due to a new process of liquid sulphur dioxide production that results from the use of waste gases of the reverberatory furnaces.

It is interesting to note that while the hypothetical heat energy requirements increase almost twofold over the actual values in Table 6.1 they can be met by heat produced in the plant's own heat recovery installations; therefore the input part of the design balance does not include heat energy. In the output part new products appear, sulphur dioxide and heat energy, and the amount of sulphuric acid produced increases. As a result, the energy efficiency of the process increases from 7.2 to 20%, i.e. by a factor of 2.8.

But a study of the energy use on the basis of energy efficency only can lead to incorrect conclusions, because the qualitative differences of the energy forms are not taken into account.

The exergy efficiency of the production process in the considered case, unlike the energy efficiency, increases from 6.0 to 11.2%, i.e., approximately by a factor of 1.9. The rather large difference between energy and exergy efficiencies is due to the heat energy among the useful products. The exergy of the heat supplied by the production process to other consumers accounts for only about 20% of its total enthalpy. The amount of heat in the useful products in energy terms amounts to 40%, and in exergy terms only to about 15%. In our opinion, the latter value more correctly shows the role of thermal secondary energy resource utilization in increasing the energy efficiency of the plant. The total amount of energy losses in

**Table 6.3** Calculated energy balance of copper production

| Balance items | Mass [t] | Energy [MWh] | Exergy [MWh] |
|---|---|---|---|
| INPUT | | | |
| Sulphide ore | 41.5 | 4.64 | 4.16 |
| Concentrates from external suppliers | 1.79 | 2.07 | 1.74 |
| Pellets | 0.41 | 0.445 | 0.375 |
| Blister copper from external suppliers | 0.255 | 0.247 | 0.195 |
| Limestone | 0.49 | 0.0 | 0.0 |
| Quartz and other ores | 0.29 | 0.05 | 0.05 |
| Coal, t c.e. | 0.75 | 6.12 | 5.93 |
| Fuel oil, t c.e. | 0.143 | 1.16 | 1.08 |
| Electricity from external suppliers | – | 2.03 | 2.03 |
| Heat from external suppliers | – | – | – |
| Energy from external suppliers, total | – | 16.762 | 15.565 |
| OUTPUT | | | |
| Cathode copper | 1.0 | 0.981 | 0.763 |
| Commercial anodes | 0.098 | 0.096 | 0.075 |
| Copper vitriol | 0.028 | 0.004 | 0.005 |
| Sulphuric acid (monohydrate) | 1.93 | 0.316 | 0.214 |
| Sulphurous anydride (liquid) | 0.43 | 0.609 | 0.415 |
| Heat of waste-heat boilers to external consumers | – | 1.347 | 0.27 |
| Useful energy, total | – | 3.353 | 1.742 |
| Energy and exergy losses | – | 13.409 | 13.823 |

$\eta_{en} = 3.353/16.762 = 0.20$; $\eta_{ex} = 1.724/15.565 = 0.112$

the energy balance will be 16 % lower than in a real production plant, although the total energy consumption will increase. Table 6.2 illustrates what types of losses change and how, provided that the chemical and thermal secondary energy resources of the production process are completely utilized.

These calculations have been confirmed experimentally, in particular, the utilization of thermal secondary energy resources and $SO_2$ trapping from the waste gases. It has been shown that production of liquid sulphur dioxide from the waste gases of reverberatory furnaces is profitable if the $SO_2$ concentration in the gas exceeds 2 %. More complete trapping of $SO_2$ from converter gases for further sulphuric acid production and reduction of their gas volume with corresponding increase of $SO_2$ concentration have a significant effect.

The economic expediency and efficiency of utilizing the thermal secondary energy resources was confirmed by calculations using the technique described in Chap. 4. The SER sources were taken to be the heat of the waste gases from reverberatory furnaces and converters. The dynamic programming method was used to calculate the minimum utilization costs and the optimum extent of SER use with an assumed total SER thermal capacity of 107 Gcal/h (124 MW). These calculations showed that with the given capacity of a cogeneration plant, given levels of electricity and heat consumption in the region accepted technico-economic indices the optimal thermal output of heat recovery installations amounted to 90 Gcal/h (105 MW). These calculations did not take into account the improvement of the air around the plant due to reduction of harmful emissions.

### 6.1.2 Autogenous Processes

As discussed above, each substance has a certain chemical energy and exergy which become apparent when the substance is used as a raw material for obtaining some product. Some substances have high chemical energy, others have low energy, even close to zero. Organic compounds typically have high values of chemical energy. Inorganic substance, for example, the metals and their sulphides that are found among compounds of processed raw material, can also have high energy content. This energy is released during the production process. If the process is designed so that the chemical energy of raw material can be used with maximum efficiency, then the consumption of conventional energy resources and energy carriers can be considerably reduced. In some cases processing of raw material can even generate conventional sources of fuels, or electrical and thermal energy. This is the primary goal of industries which consume raw materials with high chemical energy. In particular, this is the basis for the redisign of non-ferrous metallurgy, that is, autogenous processing, proceeding practically without hydrocarbon fuel.

It is not so long ago that energy aspects of industrial processes were considered only fleetingly in the course of designing new metallurgical technologies. The problems of rational energy use were not raised and solved. For this reason, a major part of the chemical energy of the raw material in many existing technologies is lost during preliminary processing before it enters the main production process.

Although efforts to increase the energy efficiency of such production processes abound, mostly they are directed at improving existing processes, rather than towards radical restructuring. The most beneficial results would be obtained if the issues of rational energy use were considered at the initial stage of technology development and would be the main design criterion. The use to autogenous processes in non-ferrous metallurgy, that is, production of non-ferrous metals by using the chemical energy of the raw material is an example of the application of this philosophy.

Various modifications of flash smelting form the prevailing group among these autogenous technologies. They are some of the most interesting pyrometallurgical processes since they meet a number of very important requirements: For exmaple,

– more complete and effective sulphur utilization, which reduces gaseous sulphur emissions into the atmosphere and increases the production of useful products such as sulphuric acid, liquid $SO_2$, and free sulphur;

– increase in the chemical reaction efficiency to limit capital investments and increase labor productivity;

– increased use of computer-aided process control and, hence, reduction of labor in the output product.

Currently several variants of flash smelting exist

1. Heated air blast smelting
2. Oxygen-flash or flame smelting
3. Oxygen-flash cyclone-electrothermal smelting.

Heated air blasting occurs in a furnace shaft; roasting and smelting of the raw material are combined in the flame of this shaft which is formed by burning the crushed and dried sulphides in the air preheated to the required temperature. This process was first applied on a commercial scale in 1948 in Finland by the Outokumpu firm at the Harjavalta Works copper smeltery.

These furnaces allow autogenous processing of concentrates of different composition, as long as their energy potential is sufficiently high, that is their chemical energy (exergy) is high.

Table 6.4 shows the energy and exergy of copper concentrates produced from ores of different fields. One sees that these values can vary over a wide range (for the compared concentrates they differ by a factor of 2.5), depending on their composition. These characteristics can be obtained by using the technique described in Chap. 2 which is based on the mineralogical composition of the raw material, and the suitability of a concentrate for autogenous flash smelting can thus be preliminarily estimated.

**Table 6.4** Dependence of chemical energy and exergy of copper ores on the composition

| Content of main components [%] | | | Energy | Exergy |
|---|---|---|---|---|
| Cu | Fe | S | [kWh/t] | |
| 34.48 | 5.36 | 13.65 | 1087 | 912 |
| 16.60 | 30.80 | 32.50 | 2300 | 2060 |
| 17.75 | 26.70 | 32.63 | 2394 | 1883 |
| 27.70 | 35.80 | 33.35 | 2716 | 2119 |
| 15.72 | 32.54 | 40.45 | 2722 | 2439 |

As before, however, in order to be able to decide whether it is more energy efficient to process the raw material by an autogenous or traditional technology the energy balances must be calculated. In the following, we discuss the energy

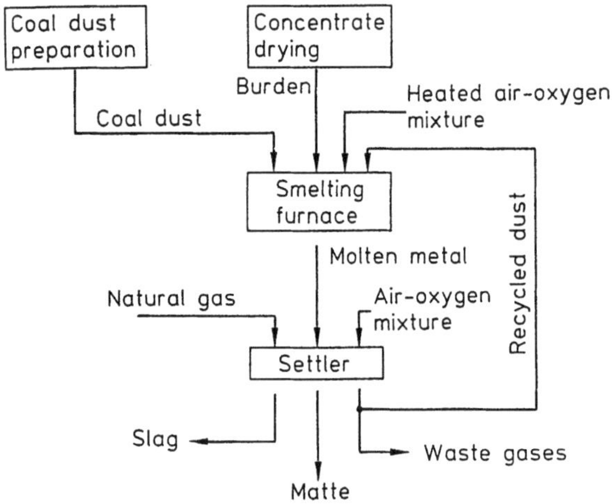


**Fig. 6.3.** Flash smelting of copper concentrates

**Table 6.5** Material and complete energy balance of a flash smelting furnace (the zone of the furnace shaft and settler) per ton of copper concentrates

| Balance items | Mass | Energy [kWh] | | Exergy [kWh] | |
|---|---|---|---|---|---|
| | [t] | $I_{ch}$ | $I_q$ | $E_{ch}$ | $E_q$ |
| INPUT | | | | | |
| Copper concentrate | 1.0 | 2716 | – | 2120 | – |
| Recycled dust | (0.215) | (500) | – | (408) | – |
| Sandstone, iron scrap | 0.43 | 330 | – | 250 | – |
| Natural gas | 0.024 | 320 | – | 300 | – |
| Oxygen | 0.123 | – | – | – | – |
| Air | 2.06 | – | 117 | – | 49 |
| Electricity | – | 69 | – | 69 | – |
| TOTAL: | 3.637 | 3435 | 117 | 2739 | 49 |
| | | 3552 | | 2788 | |
| OUTPUT | | | | | |
| Matte | 0.495 | 945* | 129* | 747* | 104* |
| Slag | 0.81 | 368 | 340 | 223 | 278 |
| Recycled dust | (0.215) | (500) | 63 | (408) | 52 |
| Waste gases | 2.32 | 790 | 900 | 540 | 740 |
| Losses | 0.012 | – | 17 | – | – |
| TOTAL: | 3.637 | 2130 | 1449 | 1510 | 1174 |
| | | 3552 | | 2684 | |

$\eta_{en} = (945 + 129)/3552 = 0.303$; $\eta_{ex} = (747 + 104)/2788 = 0.305$

*Useful energy (exergy) of the process

balance of the flash smelting with air blast (Finnish method). First, the production scheme is briefly described (Fig. 6.3).

The process is intended for flash smelting of sulphide concentrates. The charge is fed into the furnace shaft through burners. The concentrate components are oxidized by oxygen-enriched air heated to 200°C. The extent of the air oxidation is determined by the composition of the concentrate and is chosen so that an autogenous process can proceed (maximum oxidation is 40 %).

If the supplied energy cannot maintain the smelting (chemical energy of the charge drops, heat losses of the furnace grow for some reason, air temperature decreases), then natural gas is also fed into the furnace shaft.

Smelted particles of the charge emerge on the surface of a settler. The molten sulphide-silicate material is separated into layers of slag and matte in the furnace settler. To maintain the slag and matte at the required temperature the settling zone is heated by natural gas burnt in it.

At the outlet of the furnace shaft the gases change their direction by 90°, the gas flow moves horizontally over the bath in the furnace settler. Then the direction of gases changes again by 90° and gas moves along the uptake of the furnace. In the uptake, where coal dust is injected, the sulphur dioxide is reduced to free sulphur.

The smelter has a complex cooling system. Two types of cooling water are used: surface water (for cooling the uptake and furnace shaft jackets) and chemically treated water for water jackets. The smelter must be air-tight to avoid air inflows that disturb the heat balance, dilute the waste gases and increase their volume. The latter leads to excessive consumption of reducing material.

The concentrate material used in flash smelting should meet the following requirements: water content should not exceed 0.2 %, the size of 95 – 96 % of material grains must be 44 $\mu$m; the size of 90 % of the sandstone – 2 mm. Typically, the flash smelter is charged with copper concentrate, recycled dust (from the flash smelting furnace), sandstone, coal dust, industrial grade oxygen, blast air, natural gas, nitrogen, and iron scrap.

Three products are obtained:

- molten particles of the charge in the settler;
- oxidized fine dust;
- waste gases containing $SO_2$.

The main processes leading to slag formation in the settler are exchange reactions, physical separation of molten material into matte and slag due to their difference in density.

The gases evolved in flash smelting contain, before reduction, 12.6 % $SO_2$, 8.5% $H_2O$, 5.5 % $CO_2$, 0.7 % $O_2$, 72.0 % $N_2$ and their temperature is 1450°C.

The sulphide gas is reduced in the uptake of the flash smelting furnace. Fine coal, with a minimum volatile component content and minimum ash is used as a reducing agent. Carbon of coal dust interacts with sulphur dioxide, reducing it to free sulphur.

This production scheme was the basis for our calculation of its material and complete energy balance. The chemical energy and exergy of the raw material and other, non-combustible materials were calculated from their mineralogical composition using the technique described in Chap. 2. The chemical energy of fuels was assumed to be equal to their calorific value, and exergy was determined by the formulas of Rant [132].

Table 6.5 gives the material and complete energy balances of processes occurring in the furnace shaft and settler of the flash smelting furnace. The energy and exergy efficiencies of copper concentrate smelting are found to be approximately the same and are equal to $\eta_{en} = 30.3\,\%$, $\eta_{ex} = 30.5\,\%$. The ratio of energy carriers participating in the process is as follows:

for energy – $I_r : I_f : E_{el} : I_q = 0.858 : 0.090 : 0.033 : 0.019$: for exergy – $E_r : E_f : E_{el} : E_q = 0.851 : 0.108 : 0.016 : 0.025$, where $I_r$ and $E_r$; $I_f$ and $E_f$ are the chemical energy and exergy of the raw material and fuel, respectively.

Calculations have shown that the chemical energy of the raw material accounts for about 86 % of the energy supplied for copper concentrate smelting, and 9 % is contributed by other fuel. The latter is consumed mainly for maintaining the required temperature in the settler rather than directly for the smelting.

In order to determine the energy efficiency of the entire flash smelting facility, the complete energy balance was calculated for all the processes occurring in the furnace, including reduction of the waste gases to produce free sulphur (Table 6.6). $SO_2$ reduction requires an additional amount of coal dust, but the appearance of the new useful product improves the efficiency level of the raw material, consumption reduces sulphur dioxide emissions and, despite the additional costs, leads to an increase in the efficiency of the entire process; the efficiencies increase to $\eta_{en} = 39.5\,\%$, $\eta_{ex} = 38\,\%$.

The relative fraction of chemical energy (exergy) of the raw material in the ratio of energy carriers tends to decrease and the fraction of fuel tends to increase:
for energy – $I_r : I_f : E_{el} : I_q = 0.616 : 0.346 : 0.024 : 0.014$;
for exergy – $E_r : E_f : E_{el} : E_q = 0.573 : 0.398 : 0.012 : 0.017$.

Comparison of the energy efficiency of flash smelting with that of reverberatory smelting proves its obvious advantages with respect to both the efficiency and ratio of energy carriers. Possibilities for further increase in the efficiency of flash smelting by heat recovery of the waste gases are not considered here.

**Table 6.6** Material and complete energy balance of a flash smelting furnace per ton of copper concentrates

| Balance items | Mass | Energy [kWh] | | Exergy [kWh] | |
|---|---|---|---|---|---|
| | [t] | $I_{ch}$ | $I_q$ | $E_{ch}$ | $E_q$ |
| INPUT | | | | | |
| Copper concentrate | 1.0 | 2716 | – | 2120 | – |
| Recycled dust | 0.215 | (500) | – | (408) | – |
| Sandstone, iron scrap | 0.43 | 330 | – | 250 | – |
| Coal dust, nitrogen | 0.219 | 1390 | – | 1350 | – |
| Natural gas | 0.024 | 320 | – | 300 | – |
| Oxygen | 0.123 | – | – | – | – |
| Air | 2.06 | – | 117 | – | 49 |
| Electricity | – | 69 | – | 69 | – |
| TOTAL: | 4.071 | 4825 | 117 | 4089 | 49 |
| | | 4942 | | 4138 | |
| OUTPUT | | | | | |
| Matte | 0.495 | 945* | 129* | 747* | 101* |
| Slag | 0.81 | 368 | 340 | 223 | 278 |
| Recycled dust | 0.215 | (500) | 63 | (408) | 52 |
| Free sulphur | 0.152 | 879* | – | 721* | – |
| Waste gases | 2.287 | 887 | 1060 | 644 | 862 |
| Losses | 0.012 | 28 | 243 | 23 | – |
| TOTAL: | 4.071 | 3107 | 1835 | 2358 | 1269 |
| | | 4942 | | 3654 | |

$\eta_{en} = (945 + 879 + 129)/4942 = 0.395$; $\eta_{ex} = (747 + 721 + 104)/4138 = 0.38$

*Useful energy (exergy) of a product

## 6.2 Lead and Zinc Production

Lead and zinc are produced from polymetallic ores, primarily sulphides PbS and ZnS. The ores usually contain up to 1 – 2 % Pb and 2.5 % Zn.

After flotation concentration the content of the lead concentrates is 40 – 78 % lead, 5 – 10 % zinc, 3 – 4 % copper. Additional components are gold, silver, antimony, arsenic, cadmium, tin and other elements. The leftover gangue consists of pyrite, quartz, limestone, dolomite, aluminum, etc.

Zinc concentrates produced by the selective flotation of the polymetallic ores usually contain 46 – 62 % Zn, 1.2 – 2.5 % Pb, 3.5 % Cu, 0.25 % Cd, 3 – 10 % Fe and 30 – 38 % S.

After concentration, lead and zinc productions continue along separate paths. The efficiencies of these production processes are thus usually considered independently.

### 6.2.1 Production Scheme and Energy Balances in Lead Production Using Blast Smelting

Agglomeration, reduction, blast smelting and refining are the main stages of the lead production scheme based on blast smelting. They are accompanied by various intermediate processes (Fig. 6.4).

Blast smelting of agglomerate is the most commonly used method of lead metal production (about 80 % of the world lead production). It involves reducing reactions and is a continuous, highly efficient process that extracts 95 – 96 % of black lead. Blast smelting can be used for processing all kinds of ores, lead-rich and lead-poor, sulphidized and oxidized, pure and with different impurities. Coke is used as the fuel and reducing agent in the blast furnaces. Usually the air used in blasting is oxygen-enriched. This increases the furnace efficiency, reduces the waste gas temperature and lead losses, and improves the extraction rates of associated metals.

Before smelting in the blast furnaces, the lead concentrates are sinter roasted to remove sulphur which hinders the smelting process. The extent of roasting depends on the concentrate composition. If it does not contain copper, the sulphur can be removed completely. If there is copper, then the amount of sulphur after roasting should be such that the Cu content in the matte be in the range of 15 – 20 %.

Limestone and quartz are added to the furnace charge as fluxes. Quartz forms silicates that easily fuse with metallic oxides and bind the particles in the charge.

Lead, present in the agglomerate in the form of oxide or silicate, is released during smelting as a free metal. Copper oxides are first reduced to the metal and then form sulphides that are transformed into matte. If the sulphur amount in the charge is not sufficient for this reaction, some copper, reduced to metal, bonds to Pb to form black lead. Oxides of iron, calcium and silicon form the slag in the blast smelting. The products and "semiproducts" obtained in this process thus include black lead, copper-lead matte, wastes gases, and slag.

Black lead contains on the average 1 – 3 % copper, 1 – 2 % antimony, $\leq$ 1 % arsenic and tin, $\leq$ 0.03 % bismuth, and trace amounts of gold, silver, selenium, and tellurium. It is usually refined to separate out the valuable components. This includes a number of operations to remove one or several components at each step. After refining the lead is cast into pig lead and stored.

The matte, slag and waste gases are also refined further. Copper-lead matte usually contains 10 – 20 % Cu, 8 – 15 % Pb, 30 – 50 % Fe and about 20 % S. In addition, 4 – 5 % lead, $\leq$ 10 % silver and 75 % copper of the total amount contained in the charge transits to matte. Matte is collected in the settler and then is passed on for converting. About 80 % of the zinc contained in the lead ores is bound within the slag during smelting. To extract the zinc and lead from slag, they are fumed at a slag-processing plant. 95 % of the lead and 90 % of the zinc are removed to form sublimates that contain 10 – 15 % Pb and up to 60 – 70 % Zn in the form of oxides at the extent of their extraction from slag equal to 95 % and 90 %, respectively.

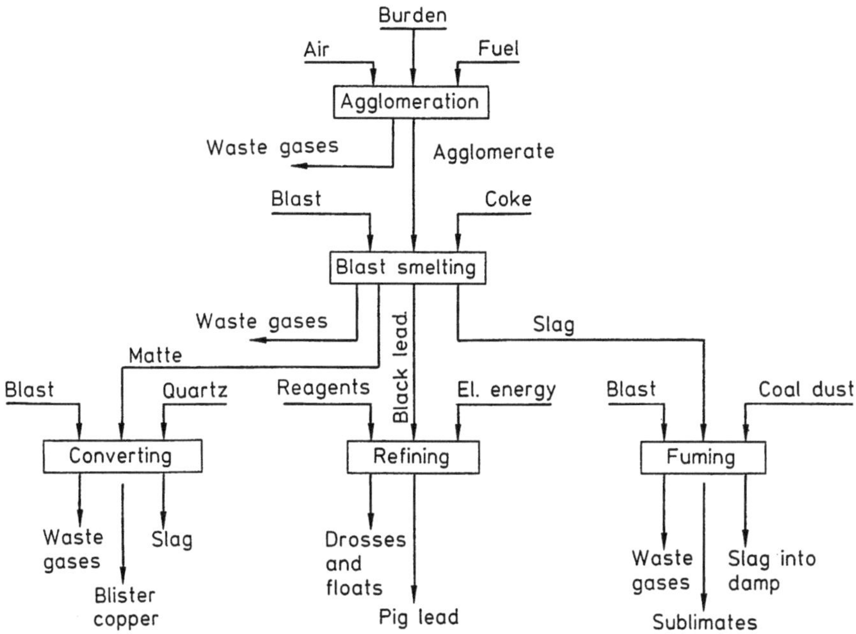


**Fig. 6.4.** Lead production using blast smelting

Large amounts of valuable components of the furncae charge are removed with dust (up to 7 – 8 % of the lead and 15 – 20 % of the zinc). Therefore, the waste gases are cleaned by filtration to trap zinc, lead, and cadmium. Thus, if the main production process is supplemented by matte and slag conversion, in addition to the zinc in sublimates and extra Pb, extra copper can be produced, whose amount depends on the copper content in the original raw material.

Sublimation fuming is a rather energy-intensive, high-temperature process producing much heat. This implies that the waste-heat boilers should be installed after the slag-processing plants for efficient heat utilization. The produced steam is consumed by auxiliary needs and is also supplied to external consumers.

The problem of sulphur utilization in lead production deserves special attention. The major amount of sulphur is removed from the original material in the form of $SO_2$ waste gas during agglomeration.

We have calculated the energy balances of a lead smeltery to determine the energy efficiency. Figure 6.5 shows the example of using the classical production scheme.

Agglomeration proceeds in agglomeration machines with air suction. Conventional energy carriers are used. Electricity is used for charge preparation and loading, driving the agglomeration machines and exhausters, air conditioning and lighting in the rooms. Oil is used in the hearths for charge ignition. Coke breeze, used as a bed for the charge placed on a travelling grate, serves as a process fuel.

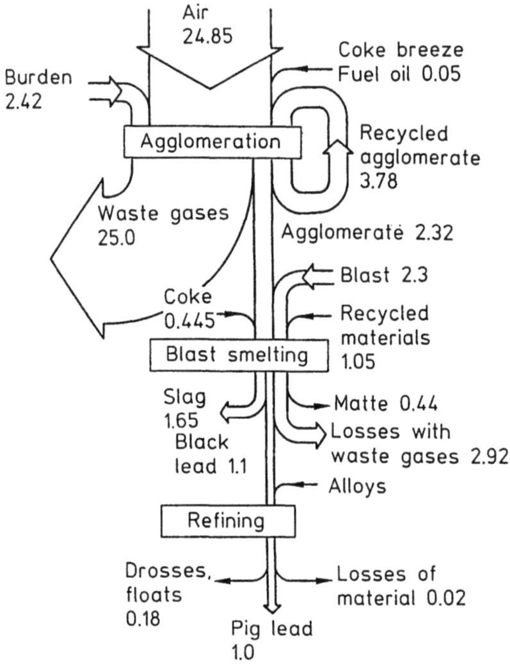


**Fig. 6.5.** Material balance of the main processes in lead production [t]

Blast smelting of the agglomerate and other lead-containing materials uses oxygen-enriched air blast. In this process, coke and electricity are used as the energy carriers. Coke is also a part of the charge and used as a fuel. Carbon monoxide is formed during burning and reduces the metals from their compounds. Electricity is used for the air pumps and material transport into the blast furnaces, for pumping the water into the water jackets, transporting the products and wastes, heating the slag in electric settlers, sucking off waste gases and dust, air conditioning and lighting the work rooms.

Black lead refining uses only electricity as an energy carrier. It is used for both heating the refineries and transporting the lead and other materials by cranes, pumping the lead by centrifugal pumps, sucking off the gases and dust, and air conditioning and lighting the rooms. In slag sublimation in the slag-processing plants, coal dust is blasted through the molten mass. Minor quantities of fuel are used for fueling the waste-heat boilers. At this stage electricity is used for driving machinery, preparing coal dust, feeding blast air and pumping cooling water for the water jackets, and for air conditioning and lighting.

Our technique was used to caclulate the complete energy balances of this lead smeltery, as a whole (Table 6.7) as well as its individual stages (the diagrams of material and energy balances are given in Figs. 6.5 and 6.6). The total energy losses at the smeltery amount to 10435.5 kWh. The contribution of the main stages to the total losses is agglomeration (2297 kWh) 22.0 %, smelting (2345 kWh) 22.5 %, refining (217 kWh) 2.0 %. Energy losses of the auxiliary stages and services account for 5576.5 kWh or 53.5 % of the total losses, the main part being due to sublimation – 4390 kWh or 78.5 %.

**Table 6.7** Energy balance of a lead smeltery

| Balance items | Mass | Energy [kWh] | | Exergy [kWh] | |
|---|---|---|---|---|---|
| | [t] | $I_{ch}$ | $I_q$ | $E_{ch}$ | $E_q$ |
| INPUT | | | | | |
| Furnace charge | 2.42 | 2280 | – | 1960 | – |
| Coke breeze, fuel oil | 0.0485 | 370 | – | 350 | – |
| Coke | 0.445 | 3400 | – | 3350 | – |
| Coal | 0.555 | 4280 | – | 4100 | – |
| Alloy | 0.16 | 55.0 | – | 40.0 | – |
| Electricity | – | 950 | – | 950 | – |
| TOTAL: | | 11335 | 10750 | | |
| OUTPUT | | | | | |
| Pig lead | 1.0 | 428 | – | 408 | – |
| Blister copper | 0.03 | 26.5 | – | 16.5 | – |
| Sublimates | 0.395 | 67.0 | – | 39.5 | – |
| Drosses, etc. | 0.18 | 88 | – | 72 | – |
| Steam of waste-heat boilers | 0.26 | – | 290 | – | 63 |
| Useful energy, total | | 609.5 | 290 | 536 | 63 |
| Energy and exergy losses | | 10435.5 | | 10151 | |

$\eta_{en} = 899.5/11335 = 0.079$; $\eta_{ex} = 599/10750 = 0.056$

The energy balance shows that in the agglomeration process the main energy losses are heat losses with the waste gases and agglomerate that account for 47 % of the total energy losses at this stage. The efficiency of this process can be somewhat increased by trapping the sulphur dioxide from the waste gases and using all the available heat of the agglomerate in the blast furnaces.

Losses of chemical energy and heat with the waste gases (58.5 %) and heat losses from the unit surfaces (41.5 %) constitute the main part of the total losses in blast smelting. The efficiency of this process can actually be increased by using the heat removed by the water in the cooling water jackets, if the water jackets are equipped with evaporative-cooling equipment.

The balance of the slag sublimation shows that this stage is rather energy-intensive. It consumes 53 % of the fuel supplied for lead production. The main losses of this process are losses with the waste gases, slag, and cooling water of the water jackets. The heat of the waste gases from the slag-processing plants is partially used in the waste-heat boilers. Currently there is no suitable technology for utilizing the heat of slags, therefore, these losses are irretrievable. Only heat removed by the cooling water can be utilized if the slag-processing plants are equipped with evaporative-cooling devices.

Analysis of the plant balance allows another important conclusion to be made. It follows from the ratio between the energy carriers used in the lead production process: $I_r : I_f : E_{el} : I_q = 0.208 : 0.717 : 0.075 : 0.0$. Only about 79.0 % of the total energy demand of the lead production is met by conventional energy

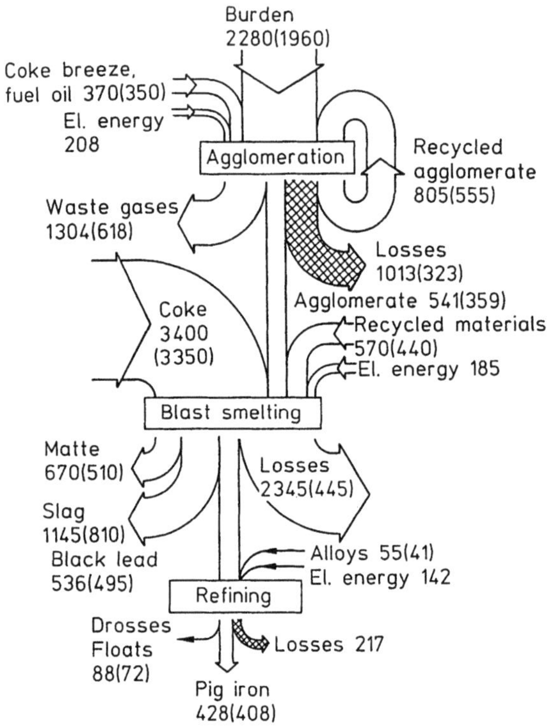


**Fig. 6.6.** Complete energy balance of the main processes of lead production [kWh] (exergy values are given in parentheses)

carriers, the remaining fraction (about 21.0 %) is contributed by the raw material. The absolute energy and exergy efficiencies of the process equal $\eta_{en} = 7.9\,\%$ and $\eta_{ex} = 5.6\,\%$.

In order to evaluate the achieved efficiency levels, as before we compare them with the maximum possible efficiencies of an idealized analog. As the analog we take lead production from pure PbS. This is a reasonable assumption, since real plants work with concentrates of which 90 % of the lead is in the form of lead sulphide. An idealized production process of only a single product (lead) has two stages: roasting and smelting. This is because chemically pure galena (PbS) is completely reduced by carbon and the refining process unnecessary.

The roasting reaction is

$$2PbS + 3O_2 = 2PbO + 2SO_2 + \Delta H_r\,.$$

The energy and exergy efficiencies of this process, $\eta_{en}^{id} = 12\,\%$ and $\eta_{ex}^{id} = 16\,\%$, can be determined by calculating its material and energy balances as was done for the copper production processes. The idealized smelting reaction is

$$2PbO + 2C + O_2 = 2Pb + 2CO_2 + \Delta H_r\,.$$

The efficiency values for this process can also be calculated in a similar way, and are equal to $\eta_{en}^{id} = 64.4\,\%$ and $\eta_{ex}^{id} = 59.6\,\%$.

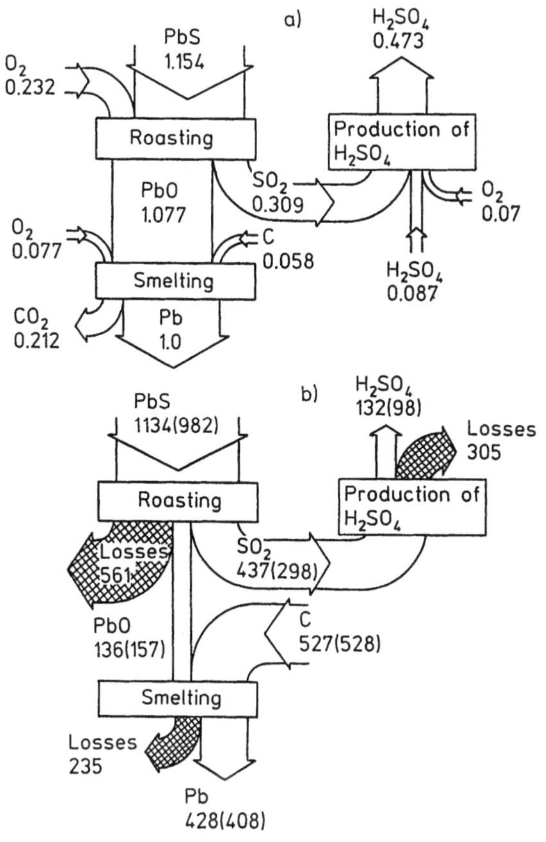


**Fig. 6.7.** Material balance [t] (*a*) and complete energy balance [kWh] (*b*) of the idealized analog of lead production using blast smelting (exergy values are given in parentheses)

These calculations allow one to determine the values of energy and exergy efficiencies for the idealized production process as a whole, $\eta_{en}^{id} = 25.8\,\%$ and $\eta_{ex}^{id} = 27.0\,\%$, provided that lead is the only useful product. Thus the relative efficiencies for the above real lead smeltery are equal to $\eta_{en}^{rel} = 30.6\,\%$ and $\eta_{ex}^{rel} = 20.7\,\%$.

At some operating lead smelteries agglomeration gases are used for sulphuric acid production. Therefore, it is interesting to observe how this new stage affects the efficiency of the idealized production. Sulphuric acid is produced via

$$2SO_2 + O_2 + 2H_2O = 2H_2SO_4 + \Delta H_r\,.$$

Since the material and energy balances of the process are known one can easily determine the efficiencies: $\eta_{en}^{id} = 30.4\,\%$ and $\eta_{ex}^{id} = 32.9\,\%$.

The material and energy balances for idealized lead production are shown in Fig. 6.7. It can be seen that utilization of sulphur dioxide from the agglomeration gases increases the energy efficiency of roasting to $\eta_{en}^{id} = 50.5\,\%$ and the exergy efficiency to $\eta_{ex}^{id} = 46.3\,\%$; the energy efficiency of the whole idealized process increases to $\eta_{en}^{id} = 33.8\,\%$ and the exergy to $\eta_{ex}^{id} = 33.5\,\%$.

As before, the individual stages of lead production can be analyzed in a similar way, potential methods for their improvement can be revealed and the impact of measures for increasing the energy efficiency of an individual process on the energy and exergy efficiencies of the entire production process can also be evaluated.

It can be shown that improvement of the energy efficiency of one process only effects the efficiency of the whole production scheme. We take as an example the agglomeration stage in lead production.

At some lead smelteries the agglomeration process involves air blasting through the charge. The amount of air used depends on the exhauster capacity, and the density and thickness of the charge. This air blasting yields large amounts of agglomeration gases containing 0.5 – 2.0 % sulphur dioxide. Trapping the $SO_2$ from these gases to produce sulphuric acid is not profitable with presently available technology.

Other lead smelteries use a more advanced method of sinter roasting, so-called blast roasting. In this method the amount of blast air can be varied, the air can be preheated and oxygen-enriched. Thus, the capacity of the agglomeration machines can be considerably increased. The volume of agglomeration gas is substantially reduced and their $SO_2$ concentration increases. This makes sulphuric acid production economically feasible.

The energy and socio-economic effect due to application of these two methods of agglomeration were compared using the material and energy balances calculated for 1 ton of pig lead shown in Figs. 6.8 and 6.9.

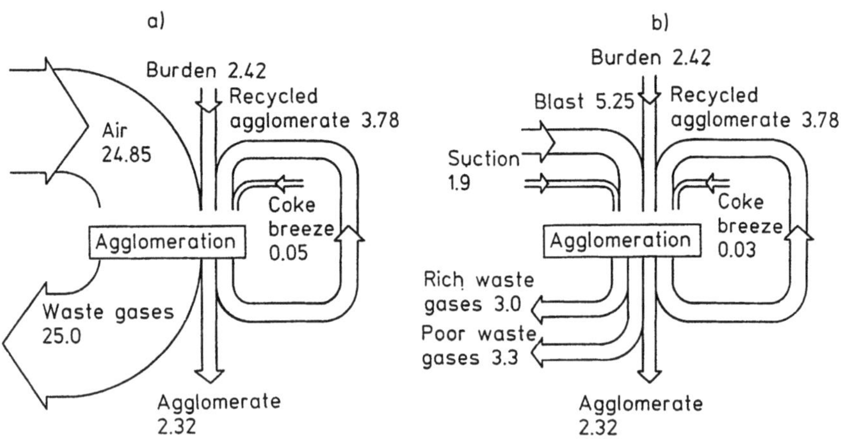


**Fig. 6.8.** Material balance of the agglomeration process [t]: (*a*) according to scheme *a*, with air suction; (*b*) with air blast, scheme *b*

Comparison of the indices for these two agglomeration schemes shows that the process using air suction (scheme *a*) requires 24.85 tons of air from the environment. 25 tons of gases at a temperature of about 100°C are produced, mean sulphur dioxide with a content of 0.8 %.

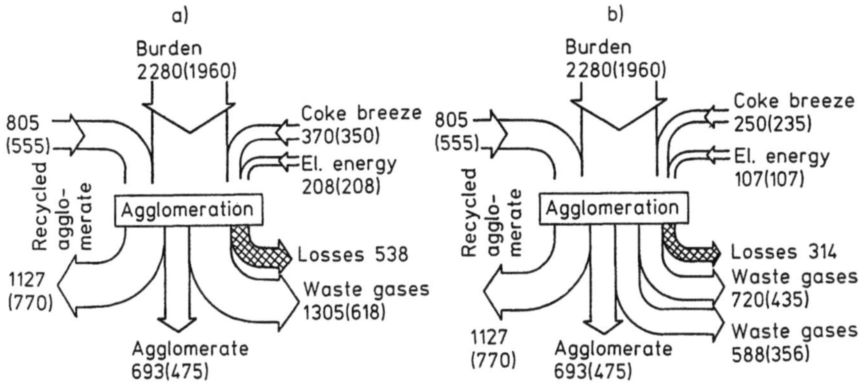


**Fig. 6.9.** Complete energy balance [kWh] (exergy values are given in parentheses); (*a*) scheme *a* with air suction; (*b*) scheme *b* with air blast

The blast-based agglomeration process (scheme *b*) requires only 5.25 tons of oxygen-enriched air and yields 6.3 tons of gases, 3 tons of which are at a temperature of about 260°C and with an $SO_2$ content of as much as 5.4 % (rich gases), and 3.3 tons at a temperature of about 400°C and $SO_2$ content as low as 2.0 % (poor gases) (Table 6.8).

**Table 6.8** Volume of evolved gases for blast-based agglomeration

| Gases | agglomeration gases | | $SO_2$ | | |
|---|---|---|---|---|---|
| | [t] | [$10^3$ $m^3$] | [t] | [$10^3$ $m^3$] | Volume [%] |
| Rich gases | 3.0 | 2.2 | 0.337 | 0.118 | 5.4 |
| Poor gases | 3.3 | 2.5 | 0.143 | 0.05 | 2.0 |
| TOTAL: | 6.3 | 4.7 | 0.48 | 0.168 | |

In scheme *a* all the agglomeration gases are cleaned from dust and emitted into the atmosphere, in scheme *b* only poor gases are emitted, rich gases are forwarded to sulphuric acid production, in which the $SO_2$ fraction is trapped out.

The air blast method allows a reduction in the energy resource consumption. This is clearly seen from the diagrams for material and energy flows of the compared processes. Due to the considerable reduction in the volume of evolved gases electricity consumption is reduced by 100 kWh, even after its additional consumption to produce oxygen for air enrichment is taken into account. In addition, the consumption of coke breeze in scheme *b* is reduced almost twofold.

When determining the efficiency of the process to the scheme *a*, usable agglomerate is the only useful product, therefore, its values of chemical anergy and exergy are in the numerator. The heat energy and exergy of the agglomerate are assumed to be partyly useful, since they are used in the next stage of the production process.

The temperature of the agglomerate put into the blast furnaces after the basin-type coolers is about 200°C, therefore, its enthalpy $I_q$ = 66 kWh and exergy $E_q$ = 27 kWh are also useful energy (exergy).

Scheme *b* yields two useful products: agglomerate and rich agglomeration gases. When calculating the efficiency for this scheme, the numerator contains the sum of the chemical energy (exergy) of the agglomerate and rich agglomeration gases and the heat energy (exergy) of the agglomerate. The heat energy (exergy) of rich agglomeration gases is neglected since it is not usefully utilized in the next stage of sulphuric acid production.

Comparison of the energy (exergy) efficiencies shows that application of the blast from below in the agglomeration machines doubles the energy efficiency and increases the exergy efficiency by a factor of 2.3.

Thus, to summarize: blast-based agglomeration both reduces the electricity consumption (for dust trapping from agglomeration gases), increases the energy (exergy) efficiency of the agglomeration process (by producing a secondary product, sulphuric acid) and reduces $SO_2$ emissions. Let us consider these components further.

**Reduction of electricity consumption.** Almost all the waste gases produced in non-ferrous metallurgy are cleaned from dust in "precipitators" and filters before being emitted into the atmosphere. This is done, first, for purely economic purposes, i.e., the trapped dust is cycled to the extraction efficiency of valuable components and, second, for environmental protection.

In our sample plant, about 39.5 tons of agglomeration gases at a temperature of about 60°C were fed from the process using scheme *a* to a dust-trapping plant. 92 kWh of electricity were consumed to clean this gas in bag filters. Hence, the electricity consumption to clean 1 ton of gas is 92/39.5 = 2.32 kWh/t or about 3 kWh per 1000 $m^3$.[1]

For comparison let us assume that the poor gases from scheme *b* are fed into the dust catching plant at the same temperature as those from scheme *a*. In practice, agglomeration gases are cooled by natural and induced suction of the air from the environment. Agglomeration via scheme *b* yields about $400/60 \times 3.3 = 2.2$ tons of gases for dust removal. Reduction of the electricity consumption for cleaning the agglomeration gases by scheme *b* will amount to $2.32 \times (39.5 - 22) = 40.5$ kWh.

**Sulphuric acid production.** Table 6.8 shows that 0.168 ton of sulphur was forwarded to sulphuric acid production with the rich agglomeration gas fraction (scheme *b*). With a sulphur extraction ratio of 92 %, 0.155 ton of sulphur, or 0.31 ton of $SO_2$, will be trapped. This amount of sulphur can produce 0.56 ton of sulphuric acid monohydrate. Therefore, transition to the agglomeration technology of scheme *b* allows 0.56 ton of sulphuric acid to be produced per ton of pig lead.

[1] These are actual data of an online plant. According to these data electricity consumption for cleaning 1000 $m^3$ of gas in the bag filters accounts for 2.0 – 2.5 kWh [46].

Note that the manufacture of a new product out of the same raw material must reduce the costs of the previously manufactured product, since a part of the costs of agglomeration, including the energy cost, will be assigned to the new product.

**Environmental protection.** Environmental problems can no longer be ignored in analyses of plant efficiency [26, 30, 46, 54, 104]. Non-ferrous metallurgy is one of the most powerful sources of environment pollution, releasing harmful gases and dust. Sulphur oxides are one of the most hazardous components. Thus agglomeration machines which can reduce $SO_2$ emissions are highly desirable.

Table 6.8 shows that only 0.142 ton of the total amount of sulphur dioxide present in the agglomeration gases occurs in poor gases. As was noted above, 0.31 ton of $SO_2$ is trapped from the rich agglomeration gases (initially containing 0.337 ton of $SO_2$) during sulphuric acid production. Hence, 0.17 ton of $SO_2$, or 35.5 % of its total amount, produced in agglomeration, is emitted into the atmosphere together with the poor agglomeration gases and waste gases of sulphuric acid production. Therefore, transition to scheme *b* allows the $SO_2$ emissions to be reduced by a factor of 2.8.

This discussion has shown that the improvement of one stage only, allows one to considerably increase the energy efficiency (twofold in the considered stage) and to achieve a notable side-effect in the other stages. This effect is determined by the reduction of electricity consumption for dust removal reduction of $SO_2$ emissions, and profit from the additional product obtained (sulphuric acid).

All these facts together greatly affect the energy efficiency of the production process as a whole. Comparison of the energy balance of the lead smeltery with the above agglomeration schemes (Tables 6.7 and 6.9) shows that the total consumption and total losses of energy are reduced. Consumption of different energy forms at a smeltery using scheme *b* is 260 kWh lower and the total losses are 375 kWh lower per ton of lead, i.e., its energy and exergy efficiencies are higher.

To show the impact of an improvement in the complexity of (multi-product) processing raw material on the energy consumption for the main product and on energy efficiency of the production process as a whole, the following comparison was made. Four variants of the technological scheme were considered. In variant 1 lead is the only product manufactured. In variant 2 the scheme is supplemented with a converting process that allows copper production. As was noted above, the volume produced depends on the copper content in the ore and concentrate. Our example considers an actual case with relatively low copper content in the ore and concentrate. Our example considers an actual case with relatively low copper content. In variant 3 a slag-processing plant is introduced into the scheme allowing zinc extraction from slag, where it appears in the course of smelting, and reextraction of lead. These calculations assume that the sublimate contains 70 % Zn and 10 % Pb.

The process of sublimation consumes much energy, therefore, in order to partially utilize the heat of the waste gases, waste-heat boilers are installed after the

**Table 6.9** Energy balance of a revised design lead smeltery

| Balance items | Mass | Energy [kWh] | | Exergy [kWh] | |
|---|---|---|---|---|---|
| | [t] | $I_{ch}$ | $I_q$ | $E_{ch}$ | $E_q$ |
| INPUT | | | | | |
| Charge | 2.42 | 2280 | – | 1960 | – |
| Coke breeze, fuel oil | 0.0325 | 250 | – | 235 | – |
| Coke | 0.445 | 3400 | – | 3350 | – |
| Coal | 0.555 | 4280 | – | 4100 | – |
| Alloy | 0.16 | 55.0 | – | 40.0 | – |
| Electricity | – | 810 | – | 810 | – |
| TOTAL: | | 11075 | – | 10495 | – |
| OUTPUT | | | | | |
| Pig lead | 1.0 | 428 | – | 408 | – |
| Blister copper | 0.03 | 26.5 | – | 16.5 | |
| Sublimates | 0.395 | 67.0 | – | 39.5 | – |
| Droses, etc. | 0.18 | 88 | – | 72 | – |
| Steam of waste-heat boilers | 0.26 | – | 290 | – | 63 |
| Sulphuric acid | 0.56 | 91.7 | – | 62 | – |
| Useful energy, total: | | 701.2 | 290 | 598 | 63 |
| | | 991.2 | | 661.0 | |
| Energy and exergy losses: | | 10083.8 | | 9834 | |

$\eta_{en} = 991.2/11075 = 0.089$; $\eta_{ex} = 661/10495 = 0.063$

slag-processing plant. Steam produced in them is also a useful product of the technological scheme.

Variant 4 suggests utilization of sulphur dioxide from the waste gases of the agglomeration stage in which the blast-based process (scheme *b*) is used. The high $SO_2$ content gases are used in sulphuric acic production.

Calculations were done for all four variants to find the exergy of the considered product relative to the total exergy of all useful products. The results of these calculations are given in Table 6.10.

The studies performed confirm the energy efficiency of the multi-product processing of raw material. Energy consumption per unit of main product (lead) is reduced by more than a factor of two with an increase in the number of useful by-products. The energy efficiency of the scheme as a whole considerably increases.

**Table 6.10** Energy consumption and efficiency for multi-product processing raw material in lead production

| Plant type (see text) | Mass | Energy consumed in [kWh] | | | | | | Total | $\eta_{en}$[%] |
|---|---|---|---|---|---|---|---|---|---|
| products | [t] | Agglomeration | Smelting | Refining | Converting | Fuming | Sulf. acid production | [kWh] | |
| *Variant 1* | | | | | | | | | |
| – lead | 1.0 | 357 | 3585 | 142 | | | | 4084 | 6.1 |
| *Variant 2* | | | | | | | | | 6.5 |
| – lead | 1.0 | 314 | 3151 | 142 | | | | 3607 | |
| – copper | 0.03 | 43 | 434 | | 45 | | | 522 | |
| *Variant 3* | | | | | | | | | 15.6 |
| – lead | 1.0 | 156 | 1567 | 142 | | | | 1865 | |
| – copper | 0.03 | 21.4 | | | 45 | | | 281.4 | |
| – zinc + lead (in sublimates) | 0.316 | 179.6 | 1803 | | | 2885 | | 4867.6 | |
| – steam of waste-heat boilers | 1.6 | | | | | 1717 | | 1717 | |
| *Variant 4* | | | | | | | | | 16.3 |
| – lead | 1.0 | 114.2 | 1567 | 142 | | | | 1823.2 | |
| – copper | 0.03 | 15.8 | 215 | | 45 | | | 275.8 | |
| – zinc + lead in sublimates | 0.316 | 131.3 | 1803 | | | 2885 | | 4819.3 | |
| – steam from waste-heat boilers | 1.6 | 78.2 | | | | 1717 | | 1717 | |
| – sulphuric acid | 0.56 | 17.5 | | | | | 53 | 70.5 | |

### 6.2.2 Zinc Production in Hydrometallurgy

Zinc concentrates derived via selective flotation of polymetallic ores are processed using pyro- and hydrometallurgical methods. The latter method accounts for about 50 % of the global zinc production.

The hydrometallurgical method of zinc production includes the following main steps (Fig. 6.10) [16, 50, 60]:

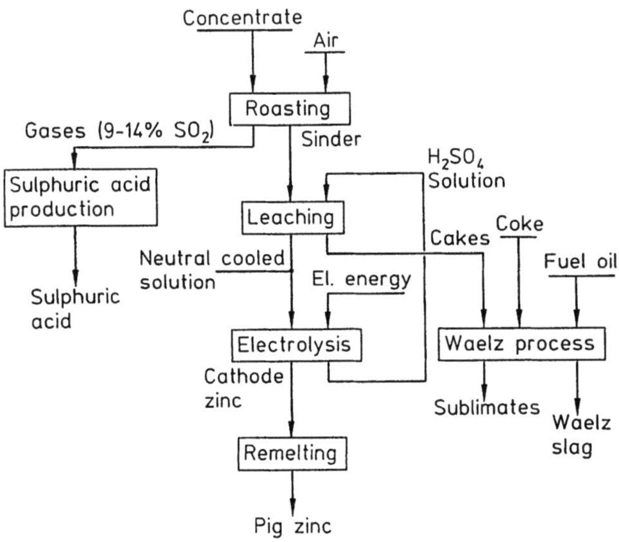


**Fig. 6.10.** Hydrometallurgical zinc production

– concentrate roasting transforming zinc sulphides to compounds soluble in sulphuric acid;

– leaching of the roasted concentrate with sulphuric acid solution;

– electrolysis of the aqueous solution of zinc sulphate;

– remelting of the cathode zinc.

In addition to the above, zinc smelteries usually have a number of auxiliary processes, e.g. $SO_2$ catchment from the roast gases for sulphuric acid production, processing of insoluble residues after leaching, and others.

**Zinc concentrate roasting.** This process, $2ZnS + 3O_2 = 2ZnO + 2SO_2$, uses fluidized bed furnaces which allow autogeneous processing, i.e., without external fuel consumption, since they use the chemical energy of the concentrate. The temperature in the furnace is maintained at 850 – 900°C. Roast gases containing 9 – 14 % $SO_2$ have approximately the same temperature at the furnace outlet, therefore, before being transferred to the sulphuric acid production area, they pass through waste-heat boilers.

**Leaching.** Roasted concentrate (cinder) is leached in aqueous sulphuric acid by $ZnO + H_2SO_4 + nH_2O = ZnSO_4 + (n + 1) H_2O$.

**Electrolysis.** Neutral solution after leaching is cleaned from impurities. Zinc is electrolyzed in basins equipped with insoluble lead or lead-silver anodes and aluminum cathodes. Zinc precipitates at the cathode, gaseous oxygen is evolved at the anode, and sulphuric acid is accumulated in the solution. The temperature in the basins is maintained at 30 – 40°C. At present, zinc smelteries use centralized cooling in vacuum-evaporating plants of the neutral solution and spent recycled electrolyte and the temperature is also regulated by the amount of solution allowed into the basins.

**Cathode zinc remelting.** The cathode zinc is formed into sheets which are remelted, in electric induction furnaces. These furnaces keep metal losses low due to their reltaively good seals and the absence of gas flows, and produce high-grade zinc. Zinc losses do not exceed 2.5 – 3 %. Electricity consumption is about 110 kWh per ton of metal. Molten zinc is cast as pigs.

As noted above, zinc smelteries produce additional by-products; most important among them is sulphuric acid from roast gases. Solid residues of leaching are reprocessed to extract zinc, copper, cadmium, lead and other elements.

**Sulphuric acid production.** Roasting zinc concentrates in fluidized bed furnaces with oxygen-enriched air blast produces gases containing 9 – 14 % sulphurdioxide. These gases pass through the waste-heat boilers installed after the fluidized bed furnaces and then go on to the sulphuric acid production step. Here the gases are first cleaned from dust and $SO_2$ is removed for the sulphuric acid production according to the contact scheme described above.

**Waelz processing of zinc cakes.** After cinder leaching there remains a solid residue – zinc cake. Depending on the composition of the input concentrates, these cakes make up 20 to 45 % of their amount. The cakes contain about 20 % of the input zinc, about half of the copper and cadmium. Lead, gold and silver are also transferred to them from the concentrate. The Waelz process is currently the most common way of refining the zinc cakes to extract these valuable components.

In the Waelz process zinc, lead and part of the cadmium are reduced from insoluble compounds, sublimated and carried away in the waste gas stream. Copper, gold, silver remain in the Waelz slag. Waste gases from the Waelz kilns are cleaned in the bag filters after preliminary cooling. The amount of zinc extracted into the sublimate is as high as 90 – 93 %, that of lead is about 90 %. Sublimates containing 50 – 65 % zinc and 10 – 20 % lead oxides are caught in the bag filter and then returned to the leaching stage.

We have calculated the energy balances of a real zinc smeltery operating in a scheme similar to the one described above. This smeltery includes all the stages shown in Fig. 6.9 and has the most efficient equipment available. Roasting occurs in fluidized bed furnaces; to increase the energy efficiency of roasting, the excess heat of the bed and the heat of the waste gases are partially utilized. Sulphur dioxide is removed from the waste gases to be used in sulphuric acid production. The smeltery includes the Waelz process and Waelz oxide leaching. The leaching is made more efficient using compressed air and steam. The smeltery uses centralized electrolyte cooling in vacuum-evaporating plants. In these facilities, some of the water evaporates because of vacuum created by steam ejectors, and cools the remaining liquor. Finally, DC current for the electrolytic baths is supplied from converting substations equipped with the silicium rectifiers which are more efficient than older mercury-arc devices.

The net energy balance for all the processes in the zinc smeltery is given in Table 6.11. The table shows the amounts of materials that supply the smeltery with various energy forms, the values of energy and exergy supplied by each material, and the total energy and exergy consumption. As is seen from the table, 47 % of al the energy needed by the zinc smeltery is supplied by the chemical energy of the raw material and only 53 % by conventional energy carriers, 16.5 % of which is fuel and 36.5 % is electricity.

**Table 6.11** Energy balance of zinc smeltery

| Balance items | Mass | Energy [kWh] | | Exergy [kWh] | |
|---|---|---|---|---|---|
| | [t] | $I_{ch}$ | $I_q$ | $E_{ch}$ | $E_q$ |
| INPUT | | | | | |
| Concentrate | 2.64 | 6100 | 0.0 | 5280 | 0.0 |
| Zinc dust | 0.05 | 80 | 0.0 | 64 | 0.0 |
| Dusts, granulated slag, from external suppliers | 0.355 | 118 | 0.0 | 78 | 0.0 |
| Coke breeze | 0.26 | 1660 | 0.0 | 1570 | 0.0 |
| Fuel oil | 0.05 | 560 | 0.0 | 525 | 0.0 |
| Electricity | – | 4900 | – | 4900 | – |
| TOTAL: | | 13418 | 0.0 | 12417 | 0.0 |
| | | 13418 | | 12417 | |
| OUTPUT | | | | | |
| Pig lead | 1.0 | 1940 | – | 1643 | – |
| Sulphuric acid (monohydrate) | 1.77 | 290 | – | 196 | – |
| Commercial Waelz slag | 0.466 | 640 | – | 580 | – |
| Cadmium | – | 35 | – | 30 | – |
| Useful energy, total: | | 2905 | 0.0 | 2449 | 0.0 |
| | | 2905 | | 2449 | |
| Energy and exergy losses: | | 10513 | | 9968 | |

$\eta_{en} = 29052/13418 = 0.216$; $\eta_{ex} = 2449/12417 = 0.197$

The smeltery does not obtain heat energy from external sources, all the auxiliary power supply is met by heat utilized in the roasting plant.

The output part of the balance shows the amounts of useful products and their energy and exergy values, the total energy and exergy losses in the smeltery as a whole, determined by the difference between the input part of the balance and the energy of the useful products. Table 6.11 also gives the net values of energy and exergy efficiencies, $\eta_{en} = 21.6\,\%$ and $\eta_{ex} = 19.7\,\%$.

Figures 6.11 and 6.12 present the material and energy balances of the main stages of zinc production. Unlike the balance of the entire production process, energy losses at individual stages are determined from in situ measurements and calculations using the proposed procedure. The values of energy and exergy efficiencies for the most energy-intensive processes are also given.

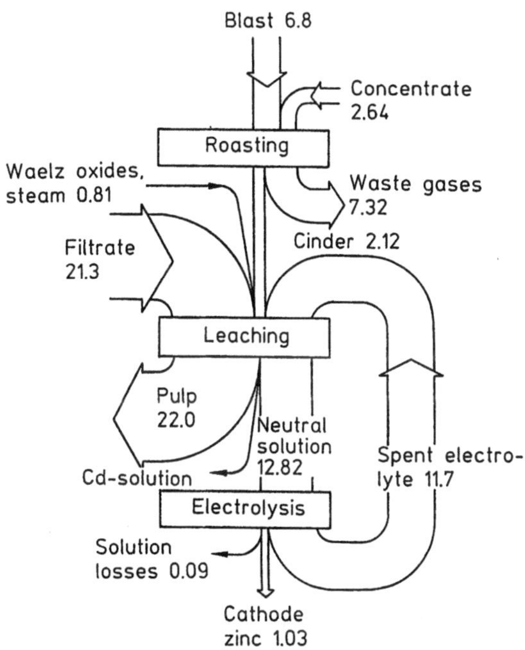


**Fig. 6.11.** Material balance of the main processes of zinc production [t]

The data show that the absolute values of energy and exergy efficiencies of the individual stages are not high. However, it is not advisable to judge the energy efficiency of the entire production process on the basis of these values only. Rather one should compare the indices of this smeltery with the best ones for an idealized analog.

The idealized analog of hydrometallurgical zinc production includes the same stages as the operating smelteries: roasting, leaching, electrolysis, sulphuric acid production.

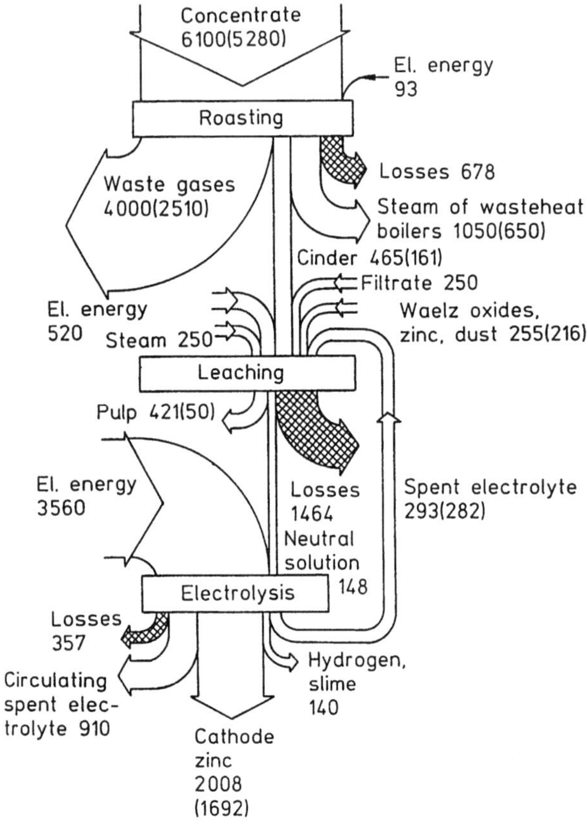


**Fig. 6.12.** Complete energy balance of the main processes of zinc production [kWh] (exergy values are given in parentheses)

Chemically pure zinc sulphide (ZnS) is used for roasting, the produced zinc oxide (ZnO) and sulphurdioxide ($SO_2$) are leached with sulphuric acid ($H_2SO_4$) and sent to sulphuric acid production, respectively.

The aqueous solution of zinc sulphate is electrolysed to yield zinc and sulphuric acid. The latter is returned to the leaching stage. Roasting and leaching are driven only by the chemical energy of ZnS and $H_2SO_4$, respectively; electrolysis requires electricity consumption.

These processes of the idealized analog are describe by the reactions

a) roasting: $ZnS + 3/2\ O_2 = ZnO + SO_2 + \Delta H_r$;

b) leaching: $ZnO + H_2SO_4 = ZnSO_4 + H_2O + \Delta H_r$;

c) electrolysis: $E_{el} + ZnSO_4 + H_2O = Zn + H_2SO_4 + 1/2\ O_2$;

d) sulphuric acid production: $SO_2 + 1/2\ O_2 + H_2O = H_2SO_4 + \Delta H_r$.

Figure 6.13 presents the material and energy flows in the idealized zinc production analog that are constructed via our calculations. The energy and exergy efficiencies for the idealized production are calculated by using the expressions

$$\eta_{en}^{id} = \left(I_{Zn} + I_{H_2SO_4}\right) / \left(I_{ZnS} + E_{el}\right) = 41.4\,\%$$

$$\eta_{ex}^{id} = \left(E_{Zn} + E_{H_2SO_4}\right) / \left(E_{ZnS} + E_{el}\right) = 39\,\%$$

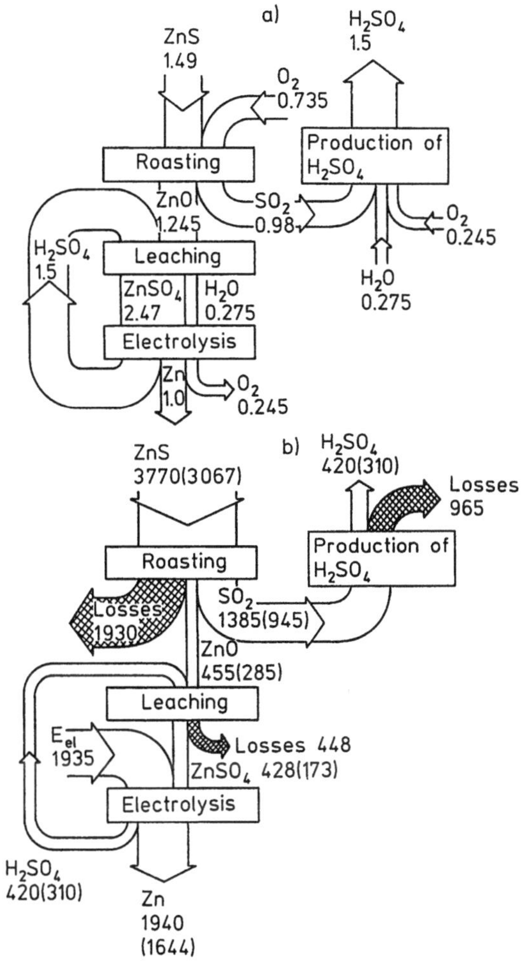


**Fig. 6.13.** Material balance [t] (*a*) and complete energy balance [kWh] (*b*) of the idealized analog of hydrometallurgical zinc production

Now the energy indices of the real and idealized production processes can be compared, i.e., the relative efficiencies are $\eta_{en}^{rel} = 52.2\,\%$ and $\eta_{ex}^{rel} = 50.5\,\%$. This energy efficiency is rather high, but it can still be increased.

The amount of energy that can be conserved and the necessary measures are again found from an analysis of energy and exergy losses. These were found from energy balances and thoroughly analyzed to reveal the possibilities for the energy loss reduction. Such a list of energy losses at the zinc smeltery is given in Table 6.12. It presents the yearly energy losses that are theoretically suitable for reuse and the unrecoverable losses in the main stages, and in the smeltery as a whole. As is seen from the table, the main fraction of the total yearly losses is due to heat losses with the roast gases (about 10 %), with cooling water and condensate of the vacuum-evaporating plants (15.3 %) and from the surfaces of equipment for sulphuric acid production (21.7 %) and leaching (12.2 %).

**Table 6.12** Primary energy losses in zinc production per ton of zinc

| Losses | Yearly energy losses [kWh] | | Exergy [kWh] | | Anergy [kWh] | | Theoretically applicable for reuse [kWh] | |
|---|---|---|---|---|---|---|---|---|
| | $I_{ch}$ | $I_q$ | $E_{ch}$ | $E_q$ | $B_{ch}$ | $B_q$ | Chemical | Heat |
| *Concentrate roasting* | | | | | | | | |
| Losses with waste gases | – | 1050 | – | 850 | – | 200 | – | 850 |
| Losses from the aggregate surfaces | – | 678 | – | 0.0 | – | 678 | – | 0.0 |
| TOTAL: | – | 1728 | – | 850 | – | 878 | – | 850 |
| *Cinder leaching* | | | | | | | | |
| Losses from surfaces and with evaporation | – | 1274 | – | 0.0 | – | 1274 | – | 0.0 |
| TOTAL: | | 1274 | – | 0.0 | – | 1274 | – | 0.0 |
| *Vaccum-evaporating plant* | | | | | | | | |
| Losses with cooling water and condensate | – | 1600 | – | 0.0 | – | 1600 | – | 0.0 |
| TOTAL: | – | 1600 | – | 0.0 | | 1600 | – | 0.0 |
| *Electrolysis* | | | | | | | | |
| Losses with hydrogen and other by-products | 140 | – | 115 | – | 25 | – | 0.0 | – |
| Losses of rectifiers and buses | – | 357 | – | 0.0 | – | 357 | – | 0.0 |
| TOTAL: | 140 | 357 | 115 | 0.0 | 25 | 357 | – | 0.0 |
| | 497 | | 115 | | 382 | | | |

| | | | | | | | | |
|---|---|---|---|---|---|---|---|---|
| *Remelting of cathode zinc* | | | | | | | | |
| Losses from surfaces and in final product cooling | – | 163 | – | 0.0 | – | 163 | – | 0.0 |
| TOTAL: | – | 163 | – | 0.0 | – | 163 | – | 0.0 |
| *Sulphuric acid production* | | | | | | | | |
| Losses with waste gases | 321 | 132 | 212 | 0.0 | 109 | 132 | 0.0 | 0.0 |
| Losses from surfaces | – | 2261 | – | 0.0 | – | 2337 | 0.0 | 0.0 |
| TOTAL: | 321 | 2393 | 212 | 0.0 | 109 | 2469 | 0.0 | 0.0 |
| | 2714 | | 212 | | 2578 | | | |
| *Filtering-drying shop* | | | | | | | | |
| Losses with wast gases | – | 260 | – | 0.0 | – | 260 | 0.0 | 0.0 |
| Losses from surfaces | – | 226 | – | 97 | – | 129 | 0.0 | 80 |
| TOTAL: | – | 486 | – | 97 | – | 389 | 0.0 | 80 |
| *Waelz processing of cakes* | | | | | | | | |
| Losses with waste gases | 108 | 380 | 100 | 232 | 8 | 148 | 100 | 190 |
| Losses with Waelz slag and sumblimates | – | 154 | – | 108 | – | 46 | – | 0.0 |
| Losses from surfaces | – | 498 | – | 230 | – | 268 | – | 190 |
| | 1140 | | 670 | | 470 | | 480 | |
| Losses of auxiliary services and shops | – | 835 | – | 0.0 | – | 835 | 0.0 | 0.0 |
| Total losses of the smeltery | 569 | 9868 | 427 | 1517 | 142 | 8427 | 100 | 1310 |
| | 10437 | | 1944 | | 8569 | | 1410 | |
| in %: | 100 | | 18.6 | | 82.1 | | 13.5 | |

It is interesting to note that if the chemical energy of raw material is disregarded in the energy balances, the main energy losses are those in the electrolysis and Waelz processes, amounting to only 4.75 % and 11 % of the total yearly losses of the smeltery.

In analyzing energy efficiency in production processes, the main emphasis should be on the real possibilities for loss reduction or recovery rather than on their absolute values. Table 6.12 shows the losses that are theoretically or actually suitable for reuse. Surprisingly, the largest potential for increasing the energy efficiency is in the concentrate roasting, although the achieved energy efficiency of this process is the highest among the production stages. This is explained by the fact that in the considered production scheme the heat energy (exergy) of the roast gases is utilized incompletely (only about 60 %). Complete utilization of the heat of roast gases in the waste-heat boilers should considerably increase the energy and exergy efficiencies of the roasting process and the smeltery as a whole. The produced additional heat energy (about 850 kWh), in the form of steam, can be supplied to external consumers. The energy efficiency of the roasting process in this case will then be as high as $\eta_{en} = (2745 + 1920)/6193 \times 100 = 75.3\,\%$ and that of the smeltery $\eta_{ex} = (2981 + 850)/13418 \times 100 = 28.6\,\%$.

The Waelz processing of zinc cakes also has considerable potential to conserve energy resources. However, this requires larger costs than improving the roasting process, where additional waste-heat boilers are needed. The energy of the waste gases and the heat lost from the external surface of the Waelz kilns could be recovered. The same is true for the heat that is lost by the external surfaces of the drier drums. The heat losses from the surfaces of drum-type furnaces are considerable due to their high temperatures.

## 6.3 Production of Titanium and Magnesium

Rutile and ilmenite are the main minerals of titanium-bearing ores. Rutile concentrate containing 95 – 97 % titanium dioxide or ilmenite (50 – 60 % $TiO_2$) concentrate is produced by the gravity concentration with magnetic separation of the ores. Titanium is usually produced together with magnesium in combined titanium-magnesium plants using a magnesium method [21, 58, 60, 119, 121, 142, 152]. The scheme of titanium-magnesium production (Fig. 6.14) has two branches, one for magnesium and one for titanium. The main stages of magnesium production are dehydrating of the starting material (carnallite $MgCl_2 \cdot KCl \cdot 6H_2O$ and bischofite $MgCl_2 \cdot 6H_2O$), electrolysis, and refining of the raw magnesium. This branch also includes the granulation of potash fertilizers and cleaning of the anode (chlorine gas) and cathode gases.

**Carnallite dehydration.** [2] To produce magnesium from chlorides the raw material – carnallite or bischofite – must be dehydrated and the iron and sulphate impuri-

[2] Bischofite is dehydrated by a similar two-stage scheme, but at lower temperatures.

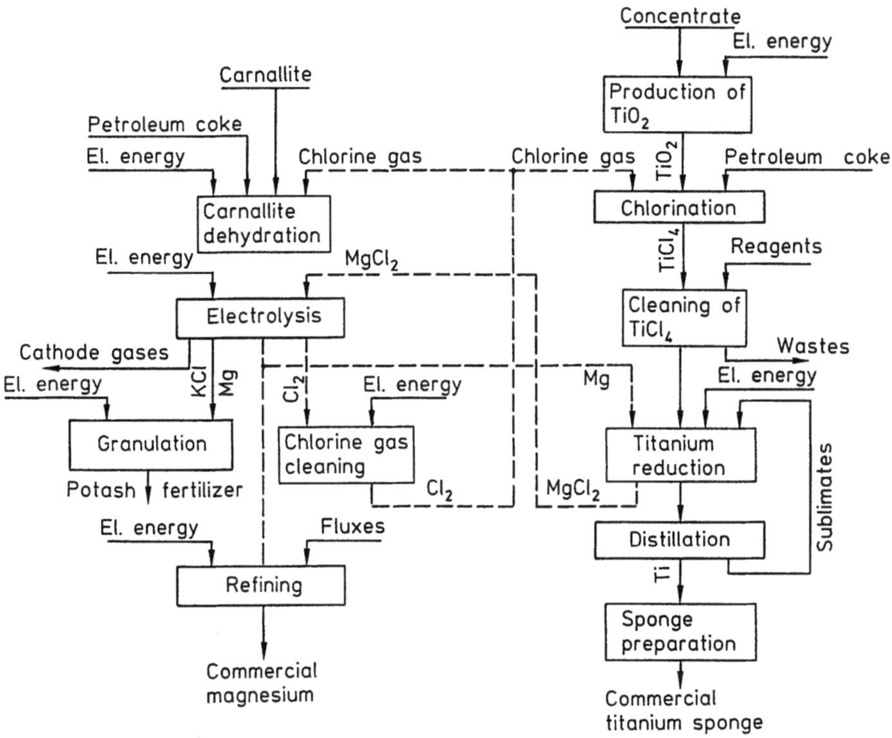


**Fig. 6.14.** Titanium-magnesium production

ties removed. Dehydration is among the most complicated processes of chemical technology and usually requires two steps. In the first, carnallite is heated in rotary pipe furnaces in the flow of flue gases whose maximum temperature does not exceed 500°C. Up to 2 – 3 % of the water is removed. The second stage of dehydration requires heating in resistive electric furnaces at a temperature of 800 – 850°C.

**Electrolysis.** In the electrolysis step, the molten electrolyte, containing $MgCl_2$ and potassium, sodium and sometimes calcium chlorides, is decomposed at a temperature of about 700°C. Electrolyzers operate without external heating. The required temperature is achieved only through the heat released in electrolysis. Metal plating out on the surface of the electrolyte is removed from the basin. Chlorine released at the anode is sucked into chlorine pipelines and transported to external consumers (chlorinators of titanium slags).

The electrolysing voltage is 6 – 6.5 V and the current density is 500 – 800 $A/m^2$. The distance between the anode and cathode is 6 – 8 cm. The coefficient of useful utilization of current is 0.8 with carnallite and 0.85 with dry magnesium chloride. The consumption of DC electric power is 17.0 – 18.0 MWh per ton of magnesium.

**Raw magnesium refining.** Raw magnesium produced by the electrolysis contains some remaining impurities (metallic and nonmetallic) that can negatively affect its corrosion resistance and mechanical properties. Na, K, Ca and Fe can appear either during electrolysis or during reduction with metallic magnesium in the original ore. Chlorides of all the components of the melt, which are trapped by the magnesium during its removal from the basin, are the main non-metallic impurities. Additional impurities are magnesium oxides, nitrides and carbides.

The magnesium is refined by either remelting with fluxes or by sublimation. Electrolytic refining of magnesium is also known.

Titanium production usually includes the following stages: production of titanium slag, chlorination of titanium slag, cleaning and reduction of $TiCl_4$, distillation and hand preparation of spongy titanium, smelting of the titanium sponge into ingots.

**Production of titanium slags.** To reduce (plate out) and separate iron and other impurities, ilmenite concentrates are smelted in elelctric furnaces. The smelting produces titanium slag containing to 90 % $TiO_2$ and iron. 95 % of the Fe contained in the concentrate ends up in the slag.

**Chlorination.** Powdered or crushed titanium slag and coke are loades into a chlorinator into the surface of molten chlorides. The chlorides are formed spent electrolyte of the magnesium electrolyzers. Chlorine from these magnesium electrolyzers is fed into the lower part of the chlorinator; it rises, mixes with the melt and reacts with it. Volatile products of chlorination are removed and condensed and the unreacted residue is also regularly removed with the melt.

**Purification of $TiCl_4$.** $TiCl_4$, with a boiling point of 136°C, is removed from the chlorinator in a steam-gas mixture into a condensation system, where it liquifies. It is partially purified since some components precipitate out as solid chlorides. $TiCl_4$ is subjected to several other cleaning stages: selective hydrolysis from the aluminum chloride, reducing oxidation from vanadium, and rectification from other impurities.

**Titanium reduction.** Titanium tetrachloride is reduced to Ti metal by heating (in electrical ovens) in an argon atmosphere at a temperature of 800 – 900°C. The reactor is loaded with magnesium (from magnesium production), closed, and filled with argon. After heating to the operating temperature, the $TiCl_4$ is put into it. The reduced metallic titanium forms a solid spongy mass soaked with magnesium and magnesium chloride.

**Distillation.** To remove the Mg and $MgCl_2$ the retort with spongy titanium mass is connected to a condenser (a similar retort) and placed into an electric furnace. Distillation is performed in vacuum at 1000°C. All the impurities are sublimated, collected and condensed in an upper condenser. After distillation, the apparatus is disassembled, the upper vessel with the sublimates is returned to reduction in the electrolysis shop and the lower vessel with the cleaned sponge is then hand sorted.

**Preparation of titanium sponge.** The cleaned titanium sponge is crushed into 12 – 70 mm pieces and sorted. The sorted sponge is either remelted or it is the final product.

**Smelting of titanium sponge into ingots.** The sponge is smelted in electric arc furnaces with a water-cooled crystallizer. The process occurs in vacuum (1.33 Pa) with melting of the consumable electrode produced by pressing the spongy titanium. The molten metal flows into a copper water-cooled crystallizer located on the bottom plate. Titanium smelting in plasma furnaces in an argon atmosphere is becoming extensively used.

The material and energy balances of a real titanium-magnesium production plant following the above scheme have been calculated. The purification of the anode gas (chlorine) and cathode gases, refining of magnesium, granulation of the spent electrolyte, are defined as auxiliary stages of the complex.

Titanium slag with 75 – 85 % $TiO_2$ and partially dehydrated carnallite (after the first dehydration stage) are used as the raw materials. The end products are titanium sponge, refined magnesium, granulated potash fertilizers, and hydrochloric acid.

The following conventional energy carriers are used in the complex: electricity for the electrolysis of magnesium chloride and carnallite; electricity and petroleum coke, for carnallite dehydration and titanium slag chlorination. Heat energy for utility services of the complex is supplied from the cogeneration plant.

**Table 6.13** Energy balance of titanium-magnesium production

| Balance items | Mass [t] | Energy [MWh] | Exergy [MWh] |
|---|---|---|---|
| INPUT | | | |
| Dehydrated carnallite | 13.13 | 2.42 | 2.19 |
| Titanium slag | 2.37 | 0.0 | 0.0 |
| Petroleum coke | 0.7 | 6.06 | 6.06 |
| Electricity | – | 54.5 | 54.5 |
| TOTAL: | 16.2 | 62.98 | 62.75 |
| OUTPUT | | | |
| Titanium sponge | 1.0 | 5.47 | 5.16 |
| Refined magnesium | 1.2 | 10.42 | 9.475 |
| Potash fertilizers | 5.0 | 0.255 | 0.1 |
| Hydrochloric acid | 0.70 | 0.228 | 0.19 |
| Useful energy, total | 7.95 | 16.373 | 14.925 |
| Losses of material, energy and exergy | 8.25 | 46.607 | 47.825 |

The material and energy balances of such a titanium-magnesium production plant are given in Table 6.13, showing the amounts of original materials and energy carriers and the corresponding values of chemical energy and exergy per ton of titanium sponge. The same quantities are given for the products. The energy and exergy efficiencies are 26 % and 23.7 %, respectively. The ratio of the energy carriers in $I_r : I_f : E_{el} : I_q = 0.038 : 0.096 : 0.866 : 0.0$. This ratio shows that the contribution of chemical energy in the raw material is relatively low and accounts for about 4 % of the total energy consumed. Electricity is the main energy carrier here; its share is more than 86 %.

We have also run a comparison of this real titanium-magnesium complex to its idealized analog. The principles governing the choice of conditions for the idealized process are the same: the raw material consists of chemically pure compounds; the products are also chemically pure; there are no material losses, reactions proceed stoichiometrically at an ambient temperature of 298 K; heat from the processes is released into the environment at the same temperature.

The idealized analog of titanium-magnesium production can be represented in the form of irreversible chemical reactions;

a) titanium slag chlorination:

$$TiO_2 + C + 2Cl_2 = TiCl_4 + CO_2 + \Delta H_r$$

b) electrolysis of magnesium chloride:

$$E_{el} + MgCl_2 = Mg + Cl_2$$

c) Reduction of $TiCl_4$:

$$TiCl_4 + 2Mg = Ti + 2MgCl_2 + \Delta H_r$$

The material and energy balances of this idealized production process show that it will consume titanium slag as raw material, electricity and carbon as energy carriers, and will produce titanium only. Hence, the values of energy and exergy efficiencies of the idealized production are equal to $\eta_{en}^{id} = 56.4\,\%$ and $\eta_{ex}^{id} = 53.3\,\%$.

The real titanium production process yields larger amounts of chlorine than stoichiometrically necessary because it is lost via leakage into the atmosphere, consumption in the initial titanium slag during chlorination, etc. Therefore, in real production plant titanium is always produced together with magnesium in some ratio.

The considered complex produces 1.2 t of refined magnesium per ton of commercial titanium sponge. This is achieved by using additional electrolyzers which operate on dehydrated carnallite.

Electrolysis of the pure dehydrated carnallite $E_{el} + KCl \cdot MgCl_2 = Mg_2 + KCl$ which produces 1.2 t of magnesium per ton of Ti, can be taken as the idealized analog of this system.

Thus both titanium and magnesium production should be considered together in an idealized analog of real titanium-magnesium production (Fig. 6.15).

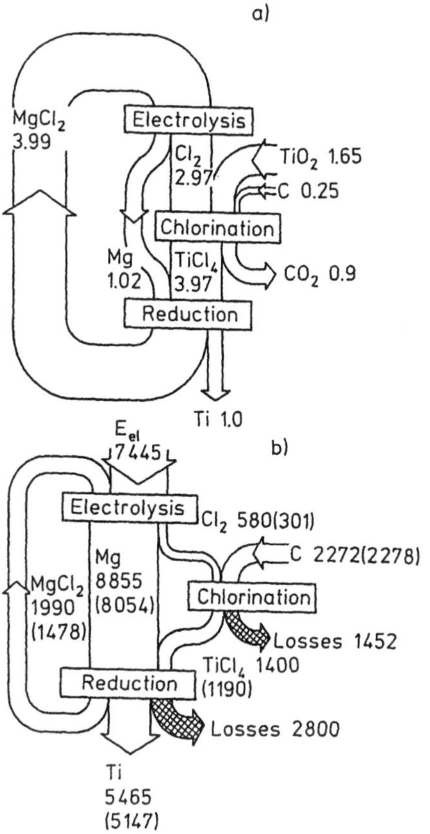


**Fig. 6.15.** Material balance [t] (*a*) and complete energy balance [kWh] (*b*) of the idealized analog of titanium-magnesium production (exergy values are given in parentheses)

The total energy consumption in an idealized production plant is 21020 kWh, 17028 KWh of which is electricity, 2272 kWh is chemical energy of carbon and 1720 kWh is the chemical energy of raw material. The idealized process has the following ratio of energy carriers: $I_r : I_f : E_{el} : I_q = 0.082 : 0.108 : 0.81 : 0.0$.

The values of energy and exergy efficiencies, provided that all the products (titanium, magnesium, chlorine and potassium chloride) are extracted and utilized, are as high as $\eta_{en}^{id} = 80\,\%$ and $\eta_{ex}^{id} = 72\,\%$. The relative efficiencies are thus $\eta_{en}^{rel} = 32.5\,\%$ and $\eta_{ex}^{rel} = 33\,\%$. These values show that the energy efficiency level in titanium and magnesium production is somewhat higher than that in copper, lead and zinc production systems considered earlier. The possibilities to increase the efficiency are relatively limited. About 95 % of the total energy losses of the complex are heat losses. Chemical energy, which accounts for 5 % of the total energy losses, is released with the slimes of electrolyzers and carnallite chlorinators, with the melts of titanium slag chlorinators and with gases.

About 50 % of all the heat losses of the complex are losses in various electrical equipment, including heat losses at a converting substation (in transformers, rectifiers), in the electrolyzer buses and in the electrical drive of machines. These losses are unrecoverable.

Heat removed with cooling water or air amounts to about 23 % of the total heat losses. This heat could be partially used by one of the currently applied schemes for utilizing low-potential losses. However, this heat is practically impossible to use due to its low temperature potential and the high aggressiveness of the gases containing chlorine and hydrogen chloride.

Calculations have revealed the usefulness of this method for the energy use analysis. Interesting and practically useful conclusions can be made on the absolute and relative efficiencies of processes, the impact of increasing multiple uses of raw material and technological improvements on the sources and causes of the largest energy losses, and which losses could be recovered as secondary energy resources. These calculations also confirmed that the method of drawing complete energy balances is much more universal and less laborious than the currently used method of heat balance. This bodes well for its prospects of further application.

# 7. Predicting Energy Conservation in an Industry by Modeling Individual Sectors

Although active, goal-oriented energy conservation policies have been pursued internationally for over ten years, development of scientific methods to design better policies is being continued even now. These methods must take into account the inter-sectoral character of energy consumption, and thus be comprehensive, using the available mathematical modeling to formalize the problem of determining the energy reserves, and the best methods for their implementation. For simplicity, in the following discussion we assume that a national economic system is under scrutiny.

The set of sectors of the economy can be considered as a composite energy-consuming system that is an independent and apparently largest block of the overall energy system of the country. The energy-consuming system has feedforwards and feedbacks, which can be internal (between subsystems) and external (to the energy supply systems). These allow one to coordinate the implemention of energy conservation measures in individual sectors, to take into account the influence of these measures in particular on the strategy of the national energy complex development. Thus, these interrelations determine the inter-sectoral character of energy conservation and necessitate the use of a complex approach. Of course, determining the energy reserves is closely related to analyzing the specific features of the production technology in each sector.

## 7.1 The Scope of the Problem

The energy reserves are defined as the difference between the energy consumption for any product produced at a given level of industrial sophistication and the future energy consumption due to more efficient equipment and technology. Thus the energy reserves are a relative quantity. Therefore, to find the energy reserves it is necessary to know the current achieved and future levels of energy consumption in the relevant technological processes. If this information is not available the reserves of energy saving can be determined only by a detailed analysis of the energy saving potentials of an individual process or technology which necessitates the use of thermodynamic methods.

The difficulty of this problem, the necessity of coordinating the results obtained for one hierarchical level for different technological processes, production schemes, or industrial sectors) with the results of other levels require a combination of different solution methods, in particular, the methods of thermodynamics and systems analysis [6, 39, 40, 47, 48, 127, 138, 145, 148, 158, 162, 165, 184, 200].

In addition, the solution method should allow one to include the effects of a variety of significant factors. One such factor is the forecasting horizon. There are at least two different cases requiring principally different approaches because of their different time horizons:

**1. 10 – 20 year Forecasts.** For this time period, one should expect a high degree of technological progress, accompanied by the introduction and incorporation of new processes and equipment. Innovations which have passed the prototype testing stage can already be included in the model since rather reliable performance, energy and economic indices are already available.

**2. Greater than 20-year Forecasts.** Such forecasts require a simplified technique since the technological and economic characteristics of processes and equipment that will operate in 20 years are generally unknown or unreliable.

In this discussion, we will be primarily considering the methods and models applied to 20-year forecasts.

The determination of energy reserves in a sector is based on the assumption that they are the sum of energy reserves in individual technological processes. These values are determined by the analysis of complete energy balance of the processes.

On the the one hand, technological progress in the industries ensures continuous improvement in the energy efficiency in each industry. On the other hand, there are also numerous factors which lead to an increase in energy consumption. These are primarily a reduction in quality of the raw materials, natural energy resources, and stricter environmental controls. These aspects usually require a lengthening in the chain of production steps, both directly within a technological scheme and indirectly, in industries which evolve to include additional processes to improve the raw materials and fuel. Concentration of poorer ores, in two stages, enrichment processing for fuels, sulphur removal and other processes of preliminary processing of coals, shales, oil might be required. In addition, new installations for treating waste discharges, and developing low-waste technologies, construction and operation of water recirculation systems can be both energy and material intensive. In the long run, all these changes lead to a growth of the energy consumption per unit of product.

Thus, only comprehensive studies including technological progress and the consequences of energy conservation policies allow the trend in future energy demands to be determined. Hence, the proposed solution technique must foresee the possibility of considering different situations and the determination of integral dependences in each case. Otherwise, in order to study the influence of a large number of different factors on industrial energy consumption, it is necessary to develop a special tool which would enable a multiple calculation of different variants for the considered industries.

One of the possible ways that this problem might be solved is to model the energy consumption system in the individual sectors with respect to measures on the rational use of consumed energy in each process. Such models are suitable for study of energy reserves in any sector. The general concepts and methodology of their formulation are stated below.

In each sector, the set of technologies, characterized by a given structure, is used for the output of some product. In the course of time this structure changes due to gradual substitution of old technologies by new ones with improved technological, economic and energy characteristics [6, 86]. This also leads to a changed energy consumption both in the absolute value and in the structure of energy carriers. The changes that have occurred in copper production may be given as an example.

This means that in forecasting technological energy consumption consideration must be given to the possible changes in the structure of technologies, since this is one of the more significant means of energy conservation. In general, the future technological structure is an optimized variable.

Production schemes of the industry are constructed in such a manner that the indices of each subsequent process depend on those of the previous one. For example, the introduction of more efficient concentration schemes and new flotation reagents can improve the coefficients of metal extraction into the concentrate. This allows a decrease in the volume of extracted ore per unit of final product (all other conditions being unchanged). An increase in the production volume in continuous casting raises the output of sound metal in rolling. This means that less steel per ton of rolled stock will be needed to be smelted which leads to a decrease in the necessary volume of extracted ore, less material to prepare, concentrate, agglomerate, etc., i.e., a change in a single link gives rise to a change along the whole technological chain.

The increasing amount of poor raw materials is particularly felt in such sectors as ferrous and non-ferrous metallurgy and the chemical industry. In some production processes this increases energy consumption, in others it may also cause technological difficulties. As a rule, the main technological process can rationally or ocess raw material of quite a certain composition with relatively small deviations of its characteristics from some optimal values. Therefore, the goal of all the preliminary processes concerning the raw material preparation consists in bringing its composition and properties to the desired level. Thus, at the natural trend of ore impoverishment in the future it should be extracted in the increasing amount, more thoroughly processed and the degree of its concentration should be raised.

On the one hand, technical progress in the mining industry leads to decrease in the specific energy consumption per unit of extracted natural raw material and, on the other hand, the trend towards reduction in the content of valuable components in ores necessitates extraction of their larger amounts at the same output of final product. At the same time energy consumption on the primary processing and concentration of ores with the decrease in their quality tends to increase. Most of the energy required for raw ore preparation is consumed in the crushing and milling stages. This depends on the physical and mechanical properties of the processed raw material. Other conditions being equal, the decrease of metal content in ore must increase the energy consumption, since to obtain the necessary coefficient of extraction, poor ores must be milled to fragments of smaller size.

Such a close interrelation between characteristics of the processed raw material and product from it, between parameters of the whole chain of processes is observed practically in all the sectors, in production of all the types of product for the whole cycle of raw material processing – from its extraction to the output of final useful product [29, 39 – 41, 152, 184]. The technique of revealing reserves of energy resource saving in the industry is developed just on the base of calculation and comparison of specific consumption values at the base and long-term levels.

Figure 7.1 shows the hierarchy of industrial sectors, products and processes in the energy consumption model. In our discussion we will elaborate on the sequence of transition from the lower levels to the upper on the example of one subsector of non-ferrous metallurgy, namely, copper production. The energy consumption per ton of Cu production is determined at the lowest level for all the possible processing methods. If the optimal contribution of each method is known, the average weighted energy consumption per unit of product can be determined. The difference in the amount of energy consumed for the base (current) and future levels represents the energy saving reserves of copper production. Similar calculations for other products, aluminum, lead, zinc, etc., can be made to find the value of the reserves of energy conserved on a national scale. Other industries are considered similarly.

In modeling energy conservation of sectors it is important to understand that the farther into the future the studied level is from the present, the larger is the change in the notion of "conservation". A variety of objective factors such as degradation in the quality of processed raw materials, improvement of product quality, stricter environmental requirements, as indicated above, can lead to an increase in the future energy consumption. However, it is important to realize that this does not mean that there is then no energy conservation effect in this case. Smaller specific consumption at the base level only indicates that the current energy requirements are lower than they will be in the future. For example, the requirements of environmental protection are at present incompletely satisfied for different reasons such as absence of finances, engineering factors, lack of necessary equipment, etc. In such cases one should emphasize that the compared variants of the sector development do not involve equal material, environmental, social and other effects.

For the comparison to be correct it is necessary to carry out calculations for the base variant that would determine the value of consumption at the present-day level, if the requirements of equal effect achieved by the existing equipment and technology were met. In such a comparison, there is always a net amount of

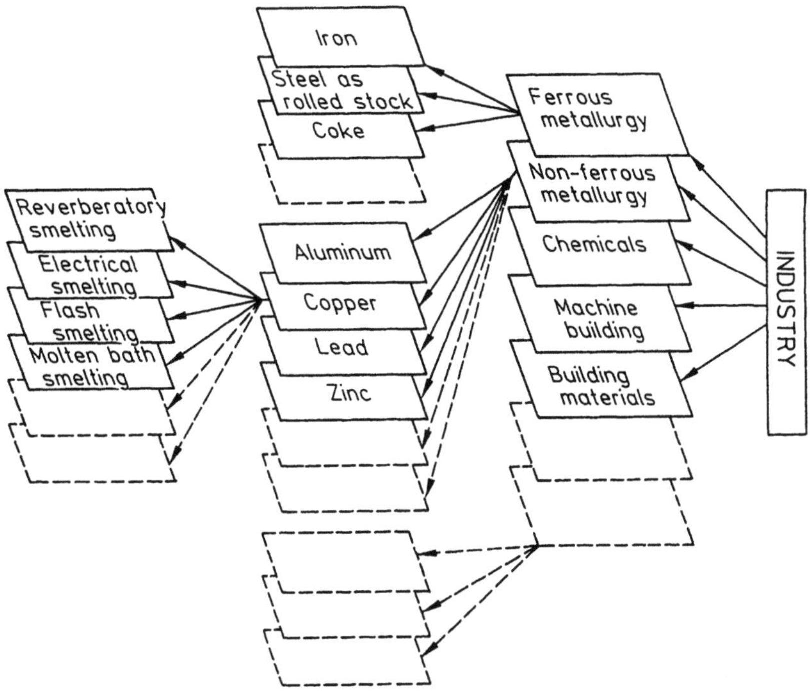


**Fig. 7.1** Hierarchy of products and processes accounted for in calculating energy consumption

energy conserved since less energy is needed to obtain the same results using the new technologies.

In this formulation, the problem of estimation of energy reserves changes into one of forecasting the energy consumption with regard for technological progress and energy conservation policies. Depending on the goal pursued, we can obtain either a vision of sector development which ensures minimum energy consumption or a definite combination of consumed energy resources (for example, with the minimum consumption of deficient coke) or, we may establish that energy consumption of a sector which corresponds to the optimal combination of technologies.

Figure 7.1 shows only the vertical production links between the levels of hierarchy within an industrial sector and the entire national economy (process – scheme – product output – sector – industry as a whole). But the horizontal links of energy consumption at the level of sectors and individual products should not be overlooked. They represent the interactions between producing and consuming sectors. For example, an improvement in a producing sector to put out product of higher quality can result in the increase of the energy consumption per unit product. However, in the sector consuming this product, an improvement in quality can lead to decreased energy consumption. Powder metallurgy, improvement of rolled stock quality of smelted metal in the ferrous metallurgy, etc. can be given as the most vivid examples of such inter-sector links. Therefore, to determine the worthiness of a given trend in technological improvement the entire system of product ouput and product consumption must be considered. Otherwise the real advantages of technological restructuring can only be determined at the level of the national economy. Thus, the true value of energy conservation in an industry can be determined only by summing these values for each sector, with regard for the interrelations between them.

## 7.2 Forecasting Energy Consumption in an Industrial Sector

A number of methods and tools have been developed to calculate future energy consumption levels. One of the ways is to plot the scheme of products and technologies for a given sector and to construct the model for forecasting its energy consumption on this basis. This allows one to evaluate the impact of each direction in technical policy and technological innovation on the sectoral energy consumption and to calculate the possible reserves of energy resource saving in comparison with the achieved level.

The main principles of devising the energy consumption model are as follows. The sector model comprises a submodel for the production of an individual product which in turn includes a submodel for the technological scheme in which its individual processes are studied (i.e., we follow the hierarchy in Fig. 7.1). The energy consumption as well as the potential amounts of conserved energy are determined at the level of this submodel, which is thus the basis of the whole calculation.

The difficulties in modeling the energy consumption of the production process are connected with the fact that the amount of energy consumed is determined by a large number of diverse parameters at the input and output from the considered system, e.g., chemical and mineralogical compositions of the raw and auxiliary materials, products and waste, physical parameters of processes, quantitative (mass) ratios of substances, etc. To model the process, i.e., to find the relations between the main input and output parameters we find the material and energy balances. The material balance is written

$$\Sigma M_r + \Sigma M_{aux} + \Sigma M_{er} = \Sigma M_p + \Sigma M_w + \Sigma M_{los} , \tag{7.1}$$

where $M_r$, $M_{aux}$, $M_{er}$, $M_p$, $M_w$, $M_{los}$ are the input and output material flows of raw and auxiliary materials, energy resources and energy carriers, useful products, waste and irretrievable losses. The general equation of the complete energy balance is given by (2.6). As was mentioned above, this balance takes into account all the energy forms used in the process, including chemical energy of the raw material and non-combustibles in the input part, and useful products, semi-products and waste in the output part. The balancing provides useful information to analyze the possibilities for improving the considered processes, since it combines the energy and exergy flows [121, 122, 151, 152, 155, 156, 159, 163, 164].

The impact of such an important factor as the change in the quality of natural raw material on the future energy consumption can also be determined by the energy balance. The quality of the raw material is measured by the content of its main components and its chemical and mineralogical composition, including barren rock. The chemical energy and exergy of raw and auxiliary materials determined from their mineralogical composition influence the consumption of traditional energy carriers and even determine the choice of the production process. For example, autogenous processes (flash smelting, smelting in a molten bath) can be used for copper ores with high specific chemical energy (exergy) while technologies with considerable external energy input (for example, electrical smelting) are required for ores with low energy content.

The energy balance is then used to determine the energy conserved due to introduction of technological measures such as an increase in pressure and temperature, or the air blast temperature. Their effects are taken into account by the recalculating the material and energy balances.

Moreover, an important advantage of the complete energy balance is the possibility of determining the total energy consumption per unit product with regard for the chemical energy of the raw and auxiliary materials. This index most correctly represents the necessary energy consumption and therefore, is more stable, easily forecasted and also enables the dependence of the consumption of conventional energy resources on the quality of raw material (its composition, chemical energy, etc.) to be found.

Frequently, forecasts of consumption of energy and energy carriers are separated into the different kinds (electric energy, heat, fuel), and then the demand for primary energy resources is determined. This results in large errors. The results obtained via our energy balance method are completely self-consistent as regards the make-up of a mixed energy supply and the total consumption. Hence, material and energy balances make it unnecessary to coordinate values of a large number of parameters that are usually contradictory, since they are obtained from different sources.

In the algorithm of the calculations, the individual processes are compiled into a technological scheme of a given product output. At the level of individual processes, provision is made for substitution of one process by another or their joint participation in the product output. All processes (both existing and proposed) whose introduction is planned during the period determined by the forecasting horizon are evaluated.

A flow chart of the algorithm for determining reserves of energy saved is shown in Fig. 7.2. For each process of the technological scheme, material and energy balances are drawn up, all the types and amounts of losses are determined, the possibility for improvements, for example, through utilization of secondary energy resources, is established. If there exists such a possibility, the balance is corrected accordingly. The achievable efficiency and the specific (per ton of product) energy consumption are determined for all the kinds of consumed energy resources (electricity, coal, coke, natural gas, fuel oil, heat). To calculate energy consumption for the considered product it is necessary to know the optimum future structure of technologies. However, its choice is a special and complex problem which will be treated below.

The cumulative effects of energy conservation at the process level on an entire sector is shown graphically in Fig. 7.3. We have an industrial sector yielding a product $i$ by produced in a technological scheme $j$ using a process $l$ which consumes different forms of energy resources. For example:

*Sector*: non-ferrous metallurgy.

*Product*: lead.

*Technological scheme*: blast smelting.

*Process*: agglomeration.

The energy consumption of the technological scheme is caclulated "backwards" i.e., first the last process that directly yields the finished product is studied. Then the previous process is calculated, etc. Consumption of each individual conventional form of energy resources (electric and heat energy, coke, natural, coke-oven and blast furnace gases, fuel oil, coal, slime, intermediate product) and the total energy consumption are determined from the energy balance of the process. The energy $\eta_{en}$ and $\eta_{ex}$ are also determined.

The following system of indices is obtained for each process:

$$\left\{\begin{array}{l} I_{ijl_1} = i_{ijl_1} M_{ijl} \\ \cdots\cdots \\ I_{ijlk} = i_{ijlk} M_{ijl} \\ \cdots\cdots \\ I_{ijlm} = i_{ijlm} M_{ijl} \\ \hline I_{ijl} = \sum\limits_{k=1}^{m} \alpha_k I_{ijlk} \end{array}\right. , \tag{7.2}$$

where $i_{ijlk}$, $I_{ijlk}$ are the specific (per unit of process $l$ product) and total (per unit of final product) consumption of energy resource $k$ for process $l$ in technology $j$; $I_{ijl}$ is the specific energy consumption of process $l$ using technology $j$ in conventional energy units; $M_{ijl}$ is the required mass of the product produced in process $l$ to obtain the unit of end product; $m$ is the number of considered energy resources; $\alpha_k$ is the conversion coefficient of energy resource $k$ in conventional energy units. Thus, both the levels of consumption of individual forms of energy resources in processes and the total energy consumption for the technological scheme are calculated.

If an intermediate product is produced via $n$ different methods in process $l$ (for example, raw material extraction can be underground or open) weighted averages are used to include all processes

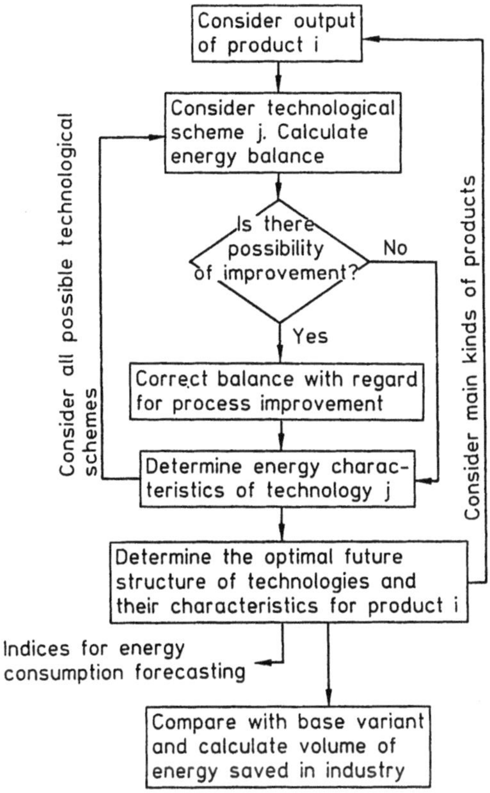


**Fig. 7.2** Algorithm for calculating the energy consumption in an industrial sector using complete energy balances of individual processes

$$\left\{\begin{array}{l} I_{ijl_1} = \sum_{p=1}^{n} i^p_{ijl_1} M_{ijl} \beta^p_{ijl} \\ \cdots\cdots\cdots \\ I_{ijlk} = \sum_{p=1}^{n} i^p_{ijlk} M_{ijl} \beta^p_{ijl} \\ \cdots\cdots\cdots \\ I_{ijlm} = \sum_{p=1}^{n} i^p_{ijlm} M_{ijl} \beta^p_{ijl} \end{array}\right. \quad (7.3)$$

where $i^p_{ijlk}$ is the specific consumption of energy resource $k$ per unit of product in process $l$ using method $p$; $\beta^p_{ijl}$ is the fraction of product $M_{ijl}$ produced via method $p$.

The values of specific consumption of all the forms of energy resources for each process of the scheme are totaled to yield the energy consumption for the technological scheme

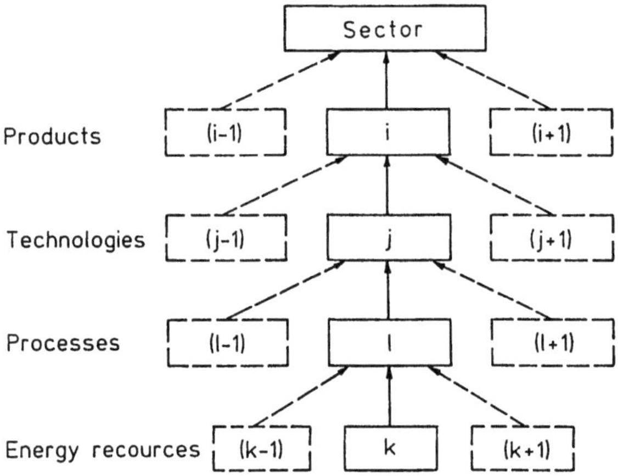


**Fig. 7.3** Cumulative energy consumption of a process-technology-product-sector scheme

$$\begin{cases} I_{ij_1} = \sum_{l=1}^{N} I_{ijl_1} \\ \cdots\cdots \\ I_{ijk} = \sum_{l=1}^{N} I_{ijlk} \\ \cdots\cdots \\ I_{ijm} = \sum_{l=1}^{N} I_{ijlm} \end{cases} \tag{7.4}$$

where $N$ is the number of processes in the scheme.

The total energy consumption on the output of a unit of product by technological scheme $j$ is

$$I_{ij} = \sum_{k=1}^{m} \alpha_k \sum_{l=1}^{N} I_{ijlk} . \tag{7.5}$$

By varying the fraction $\beta_{ij}$ it is possible to determine the weighted average specific consumption of each form of energy resource:

$$\begin{cases} I_{i_1} = \sum_{j=1}^{d} \beta_{ij} \sum_{l=1}^{N} I_{ijl_1} \\ \cdots\cdots \\ I_{ik} = \sum_{j=1}^{d} \beta_{ij} \sum_{l=1}^{N} I_{ijlk} \\ \cdots\cdots \\ I_{im} = \sum_{j=1}^{d} \beta_{ij} \sum_{l=1}^{N} I_{ijlm} \end{cases} \tag{7.6}$$

and the total average energy consumption per unit of product for the sector

$$I_i = \sum_{k=1}^{m} \alpha_k I_{ik} . \tag{7.7}$$

Here $d$ is the number of the considered technological schemes.

The difference in the energy consumption for the output of product $i$ for the current (base) and future levels, multiplied by the volume of its future output $M_i^{\mathrm{fut}}$, determines the reserves of energy resource saved

$$R_i = \left(I_i^{\mathrm{b}} - I_i^{\mathrm{fut}}\right) M_i^{\mathrm{fut}} \tag{7.8}$$

The amount of energy resources saved in the entire sector is given by the energy conserved in each subsector:

$$R_{\mathrm{sec}} = \sum_i R_i = \sum_i \left(I_i^{\mathrm{b}} - I_i^{\mathrm{fut}}\right) M_i^{\mathrm{fut}} . \tag{7.9}$$

This approach to predictions of energy consumption levels requires numerous computations of many different parameters. In addition, a broad informational base is needed which gives an idea on the trends in technological progress and also, the development of the raw material base. As was mentioned above, the quality and composition of raw material influence to a great extent on the consumption of energy resources for the end product. In some sectors this factor substantially changes in the course of time.

Moreover, it is necessary to know the technological and economic indices for the current by available and proposed technologies, and the time and possible extent to which they will be introduced into a production scheme.

Having found the initial future and current energy consumptions, we can now set about optimizing the production schemes and their substeps to try to find the maximum possible amount of energy which can be conserved. We start with a set of initial data for each process such as the degree of extracting the desired product and the amount of waste produced.

By inserting the objective function into the model and minimizing it the model becomes an optimization problem and provides a unique solution which is optimal with respect to the given criterion.

The value of the specific energy consumption in production of product $i$ can be taken as the criterion of optimality, i.e.,

$$I_i = \sum_{k=1}^{m} \alpha_k \sum_{j=1}^{d} \beta_{ij} \sum_{l=1}^{N} I_{ijlk} \rightarrow \min . \tag{7.10}$$

The minimum of $I_i$ is the optimal solution obtained for a precise technological structure of this product output and a set of energy conservation measures. This approach calculates the energy minimum in terms of production volumes.

Such a statement of the problem allows one to determine the energy minimum that is needed for the sector to realize its goal, i.e., production volumes of energy-intensive products. The material and financial expenditures required for achieving this minimum are neglected. Thus, the obtained solution is the technically possible range of a decrease of energy consumption.

Usually the minimum of real monetary costs is the generally accepted criterion of optimality, that is, when at the optimal future level of energy consumption the national-economic costs on product $i$ are minimal:

$$\sum Z_i \rightarrow \min \ .$$

The optimal technological structure for the output of product $i$ should correspond to this criterion.

However, in this formulation the problem becomes one of choosing the optimal development of a sector, which is broader in scope than the problem of devising a model of energy consumption. The former is related to the forecasting of technological progress, but is beyond the scope of the authors's competence and should be solved within a large number of constraints imposed by industrial technologists rather than power engineers.

Thus, the specific level of energy consumption corresponds to a set of initial data and, first of all, to a set of considered technologies and energy conservation measures. In other words, each level of energy consumption has quite definite material and financial expenditures, with the primary influence coming from energy conservation.

In addition to the technical information, in the data base, actual investment $k_{ijl}$, and annual costs $c_{ijl}$ (excluding energy costs) must be made available. Isolation of the energy cost component allows a quantitative evaluation of its contribution to the total costs. The calculations of these costs follow the same scheme as those for energy/exergy consumption.

The consumption coefficient $M_{ijl}$ is the mass of a product produced in process $l$, that is needed to produce a unit of end product. Multiplying $Z$, $K$, and $C$ by $M$ gives the monetary criterion of optimality, capital investment and annual costs per unit end product:

$$\begin{aligned} K_{ijl} &= k_{ijl} M_{ijl} \\ C_{ijl} &= c_{ijl} M_{ijl} + \sum_{k=1}^{m} I_{ijlk} z_k \\ Z_{ijl} &= \left(\varepsilon_s k_{ijl} + c_{ijl}\right) M_{ijl} + \sum_{k=1}^{m} I_{ijlk} z_k \end{aligned} \tag{7.11}$$

where $z_k$ are the marginal costs of energy resource $k$.

When several different processes are tested in one link of a technological chain, the weighted average $K$, $C$, and $Z$ (in accordance with the fractions of product produced by each technological scheme) are calculated via

$$\begin{aligned} K_{ijl} &= \sum_{p=1}^{n} \beta^p_{ijl} k^p_{ijl} M_{ijl} \\ C_{ijl} &= \sum_{p=1}^{n} \beta^p_{ijl} c_{ijl} M_{ijl} + \sum_{p=1}^{n} \beta^p_{ijl} \sum_{k=1}^{m} i^p_{ijlk} M_{ijl} z_k \\ Z_{ijl} &= \sum_{p=1}^{n} \beta^p_{ijl} \left(\varepsilon_s k^p_{ijl} + c^p_{ijl}\right) M_{ijl} + \sum_{p=1}^{n} \beta^p_{ijl} \sum_{k=1}^{m} i^p_{ijlk} M_{ijl} z_k \ . \end{aligned} \tag{7.12}$$

The costs for the entire technological scheme are

$$Z_{ij} = \varepsilon_s \sum_{l=1}^{N} K_{ijl} + \sum_{l=1}^{N} C_{ijl}$$
$$Z_{ij} = \sum_{l=1}^{N} Z_{ijl} \,. \tag{7.13}$$

Inclusion of additional energy conservation measures may generate additional costs, which are directly added to the costs of the technological scheme.

## 7.3 Forecasting Exergy Expenditures

Modeling the exergy consumption for an industrial sector with our algorithm by using the data of the complete energy balance is even easier than finding the energy consumption, since it is not necessary to differentiate between different forms of energy. The total exergy consumption of a technological scheme is simply the sum of the exergies of each process in the scheme, and so on up the hierarchical ladder.

The exergies can be determined from the complete energy balance by

$$\left\{\begin{array}{l} E_{ijl_1} = e_{ijl_1} M_{ijl} \\ \quad \cdot \;\cdot \;\cdot \;\cdot \;\cdot \;\cdot \\ E_{ijlk} = e_{ijlk} M_{ijl} \\ \quad \cdot \;\cdot \;\cdot \;\cdot \;\cdot \;\cdot \\ E_{ijlm} = e_{ijlm} M_{ijl} \\ \hline E_{ijl} = \sum_{k=1}^{m} E_{ijlk} \end{array}\right. \tag{7.14}$$

where $e_{ijlk}$, $E_{ijlk}$ are the specific (per unit of process $l$ product) and total (per unit of final product) exergy consumption for an energy resource $k$ in process $l$ in technological scheme $j$. The exergy consumption for all energy resources and energy carriers is reported in common units, kilowatt-hours (kWh).

The average weighted values

$$\left\{\begin{array}{l} E_{ijl_1} = \sum_{p=1}^{n} e^{p}_{ijl_1} M_{ijl} \beta^{p}_{ijl} \\ \quad \cdot \;\cdot \;\cdot \;\cdot \;\cdot \;\cdot \;\cdot \;\cdot \\ E_{ijlk} = \sum_{p=1}^{n} e^{p}_{ijlk} M_{ijl} \beta^{p}_{ijl} \\ \quad \cdot \;\cdot \;\cdot \;\cdot \;\cdot \;\cdot \;\cdot \;\cdot \\ E_{ijlm} = \sum_{p=1}^{n} e^{p}_{ijlm} M_{ijl} \beta^{p}_{ijl} \end{array}\right. \tag{7.15}$$

must be calculated when in some link of the considered chain of processes more than one process are run in parallel. The total exergy consumption per unit end product for a technological scheme $j$ is determined from the expressions

$$\left\{\begin{array}{l} E_{ij_1} = \sum_{l=1}^{N} E_{ijl_1} \\ \cdot\ \cdot\ \cdot\ \cdot\ \cdot\ \cdot \\ E_{ijk} = \sum_{l=1}^{N} E_{ijlk} \\ \cdot\ \cdot\ \cdot\ \cdot\ \cdot\ \cdot \\ E_{ijm} = \sum_{l=1}^{N} E_{ijlm} \\ \hline E_{ij} = \sum_{l=1}^{N} E_{ijk} \end{array}\right. \tag{7.16}$$

The average exergy consumption for product $i$ in a sector comprising a given structure of technological schemes is calculated as follows:

$$\left\{\begin{array}{l} E_{i_1} = \sum_{j=1}^{d} \beta_{ij} \sum_{l=1}^{N} E_{ijl_1} \\ \cdot\ \cdot\ \cdot\ \cdot\ \cdot\ \cdot\ \cdot \\ E_{ik} = \sum_{j=1}^{d} \beta_{ij} \sum_{l=1}^{N} E_{ijlk} \\ \cdot\ \cdot\ \cdot\ \cdot\ \cdot\ \cdot\ \cdot \\ E_{im} = \sum_{j=1}^{d} \beta_{ij} \sum_{l=1}^{N} E_{ijlm} \\ \hline E_i = \sum_{k=1}^{m} E_{ik} \end{array}\right. \tag{7.17}$$

The absolute value of the amount of exergy conserved in a sector is the sum of the amounts of exergy saved in all products in the sector

$$\Delta E_{\mathrm{sec}} = \sum_{i} \left(E_i^{\mathrm{b}} - E_i^{\mathrm{fut}}\right) M_i^{\mathrm{fut}} \tag{7.18}$$

The joint forecast of the consumption of energy and exergy determines the net amount of energy required and estimates of its availability needed for realization of production processes, i.e. balances the forecast with respect to energy quantity and quality.

## 7.4 Financial and Energy Expenditures for Environmental Protection

As was shown above, in most industrial sectors, output of desirable products is accompanied by simultaneous ejection of chemical substances, such as dust, gases, solid and liquid materials that for some reason do not find any useful application. Practically always, such discharges have a negative impact on the environment [26, 54, 104, 146, 168, 203].

It is recognized that the measures taken to prevent (or at least limit) the discharge of pollutants are obviously insufficient. Despite dust filters and other methods of reducing emissions, the amount of discharged waste, gaseous and condensed phase, is still rather high. Increase in production volume will of course increase waste volume. At the same time, restructions on waste emissions will become more severe. Therefore, it will be necessary to devise new and better filtering devices and waste treatment systems. But of course, this, too, requires considerable material and energy resources. The future increase of energy consumption due to implementation of environmental protection measures should be taken into account in forecasting energy consumption and determining the potential energy saving reserves.

Thus, we introduce an additional block into the model, which calculates the energy consumption for treatment of waste gases and sewage water in each technological scheme. Processing of solid wastes requires construction of another technological complex, and so, is not considered in this preliminary model. We assume that the discharges of aqueous and gaseous wastes are considerable, however. Information on the discharges of each step of the processing is transferred to this block after calculation of the corresponding material balance, which determines the volume and composition of substances removed to the environment.

As in every production process, there is always more than one way to achieve a set goal – but we need to find the minimum energy path. For example, dust filtering can be done in three (or more) stages to remove coarse, medium, and fine-grained particles. The amount of material and energy needed to build and operate the filtering system has to balance out the net environmental gain of dust removal.

In this discussion we do not attempt to design the best filtering system, but rather only demonstrate the process of calculating the energy balances for an already existing, already optimized, device.

We assume that there exists a program and a set of devices for each production process to treat its waste gases and sewage water. These devices are the most efficient possible and are characterized by a trapping coefficient, costs ($z^{\mathrm{gas}}_{ijl}$, $z^{\mathrm{sew}}_{ijl}$ for gas and sewage, respectively), and energy consumption ($i^{\mathrm{gas}}_{ijl}$, $i^{\mathrm{sew}}_{ijl}$) per unit of end product of process $l$. A given amount of waste that needs treatment (output of waste gases $v^{\mathrm{gas}}_{ijl}$, sewage water $v^{\mathrm{sew}}_{ijl}$) is formed in each process per unit of product. Their value per unit of end product is

$$\begin{aligned} V^{\mathrm{gas}}_{ijl} &= v^{\mathrm{gas}}_{ijl} M_{ijl} \\ V^{\mathrm{sew}}_{ijl} &= v^{\mathrm{sew}}_{ijl} M_{ijl} \,. \end{aligned} \tag{7.19}$$

Then, the consumption of energy resources $k$ for the treatment of the waste of process $l$ is calculated by

$$\begin{aligned} I^{\mathrm{gas}}_{ijlk} &= i^{\mathrm{gas}}_{ijlk} V^{\mathrm{gas}}_{ijl} \\ I^{\mathrm{sew}}_{ijlk} &= i^{\mathrm{sew}}_{ijlk} V^{\mathrm{sew}}_{ijl} \,. \end{aligned} \tag{7.20}$$

The total energy consumption for preventing discharges of process $l$ is determined by the expressions

$$\begin{aligned} I_{ijl}^{\text{gas}} &= \sum_{k=1}^{m} \alpha_k i_{ijlk}^{\text{gas}} V_{ijl}^{\text{gas}} \\ I_{ijl}^{\text{sew}} &= \sum_{k=1}^{m} \alpha_k i_{ijlk}^{\text{sew}} V_{ijl}^{\text{sew}} \end{aligned} \tag{7.21}$$

The costs of waste treatment for process $l$ are equal to:

$$\begin{aligned} Z_{ijl}^{\text{gas}} &= z_{ijl}^{\text{gas}} V_{ijl}^{\text{gas}} \\ Z_{ijl}^{\text{sew}} &= z_{ijl}^{\text{sew}} V_{ijl}^{\text{sew}} \,. \end{aligned} \tag{7.22}$$

For technological scheme $j$ the energy consumption is determined by summing up the values of specific consumption for all the processes $l$ of the scheme; for an energy source $k$

$$\begin{aligned} I_{ijk}^{\text{gas}} &= \sum_{l=1}^{N} I_{ijlk}^{\text{gas}} \\ I_{ijk}^{\text{sew}} &= \sum_{l=1}^{N} I_{ijlk}^{\text{sew}} \,, \end{aligned} \tag{7.23}$$

for energy of all forms

$$\begin{aligned} I_{ij}^{\text{gas}} &= \sum_{l=1}^{N} I_{ijl}^{\text{gas}} \\ I_{ij}^{\text{sew}} &= \sum_{l=1}^{N} I_{ijl}^{\text{sew}} \,. \end{aligned} \tag{7.24}$$

The material consumption for the treatment of discharges for an entire technological scheme is determined similarly, as the sum of materials consumed in each process

$$\begin{aligned} Z_{ij}^{\text{gas}} &= \sum_{l=1}^{N} Z_{ijl}^{\text{gas}} \\ Z_{ij}^{\text{sew}} &= \sum_{l=1}^{N} Z_{ijl}^{\text{sew}} \,. \end{aligned} \tag{7.25}$$

These values are then added to the values of energy consumption.

The material balances for the systems of cleaning waste gases and sewage water are drawn up on the basis of material balances of the individual production processes. They include the total quantity of process waste, the amount of trapped pollutants, and the residual quantity of emitted pollutants. Once these are known, further decisions on the design of technological schemes, and consequently, industrial sectors, can be made.

# 8. Evaluation of Energy Reserves as a Result of Energy Conservation. Ferrous Metallurgy

This chapter deals with the results of experimental calculations performed by our technique for evaluating the reserves of energy saved in ferrous metallurgy. We assume that future energy consumption will depend on the introduction of more efficient production steps and improvement o technological structure of product output, improvement in equipment, and effective utilization of secondary energy resources. Specifically, we anticipate the following large-scale energy saving measures:

a) change in the production structure of steel smelting methods;
b) introduction of continuous steel casting;
c) increase in the extent of utilization of the potential energy of the blast-furnace gas, the chemical energy of converter gas, and the heat of hot coke;
d) intensification and improvement of existing processes (e.g., increase of blast temperature and rate of oxygen feeding, heating of scrap loaded in the steel smelting furnaces).

The reserves of energy conserved were determined for some future level (level II) with respect to a ground state level (level I) corresponding approximately to its state in 1985.

Two sets of calculations were performed. In the first, the energy saving reserves were estimated for individual processes. The complete detailed energy balances were found, for such processes as agglomeration, blast-furnace process, steelmaking, production of coke and rolled stock.

The second set of calculations aimed at obtaining trends in the energy consumption as a function of various factors, for example, the ratio of the steelmaking technologies, the volume of steel involved in continuous casting, the fraction of dry coke quenching and production of pellets.

## 8.1 Steelmaking

### 8.1.1 Energy Conservation due to Technological Restructuring

One production process which should yield to a drop in energy consumption is the use of an "oxygen converter". This technology is already finding increased use [4, 17, 18, 41, 86, 93, 96, 107, 191, 204]. The advantages of the oxygen converter process, in particular, over the open-hearth furnace, are

a) higher output;
b) steel of comparable quality (rimming, killed, semikilled steels as well as a number of alloy steels: transformer, dynamo, wire-rope, tool, can be smelted)
c) automation, allowing, precision control of the pouring bay and rolling mills;
d) more rapid construction.

In the USSR, it has been shown that the cost of converter steel is 2.7 % lower than open-hearth steel, the required capital investment in a plant and its allied industries are 1.6 % lower, and productivity is 1.5 times higher [107].

Three main technological schemes are used for steelmaking in the USSR,
agglomeration – blast-furnace – open-hearth;
agglomeration – blast-furnace process – oxygen converter;
electric scrap smelting.

These schemes are characterized by different specific consumptions of electric and heat energy, fuel, including coke. Therefore, changing the relative amounts of steel produced by each method can substantially influence the energy consumption level and the kind of the most used energy carriers. Thus, even without changing the nature of the individual production steps (by technique or equipment modifications, for example) the total energy consumption by the ferrous metallurgy industry can be changed.

If possible changes in the value and structure of consumed energy resources in the technological processes due to realization of different energy conservation measures are taken into account, the difference between the average indices for the industry determines the value of total energy saving reserve for the whole complex of the considered directions in energy conservation.

Exactly such a calculation was performed. The energy balances for each scheme were drawn up and the energy consumption calculated. All the results were normalized to the known average values which give energy consumed per ton of final steel product. The energy consumption in the primary steps of each scheme is given in Table 8.1 for the different energy types.

In Table 8.2 we have listed the results of calculations for two different combinations of schemes 1, 2, and 3. As can be seen, quite a difference in energy consumption can be achieved. In particular, for the conventional energy forms 114.8 kg coal equivalent (c.e.) of fuel (including 50 kg of coke) and 1.4 kg c.e. of heat energy can be saved. Electrical energy consumption rose by 116.4 kWh per

**Table 8.1** Consumption of energy resources for different schemes of steel production

| Technological scheme, process | Product mass by material balance [t] | Energy [kWh/t] | | | | | | |
|---|---|---|---|---|---|---|---|---|
| | | Chemical energy of raw material | Chemical energy of fuel | | Electric | Heat | Total consumption | |
| | | | Total | Including coke, electrodes | | | Total | Including conventional forms |
| *Scheme 1*: | | | | | | | | |
| agglomeration | 0.91 | 166 | 505 | 435 | 32 | 18 | 721 | 555 |
| blast-furnace process | 0.542 | 109 | 3320 | 2215 | 6 | 38 | 3473 | 3364 |
| open-hearth process | 1.0 | 1032 | 1569 | – | 13 | 35 | 2649 | 1617 |
| TOTAL: | | 1307 | 5394 | 2690 | 51 | 91 | 6843 | 5536 |
| *Scheme 2*: | | | | | | | | |
| agglomeration | 1.225 | 224 | 680 | 585 | 43 | 24 | 971 | 747 |
| blast-furnace process | 0.728 | 147 | 4469 | 2980 | 8 | 51 | 4675 | 4528 |
| oxygen steelmaking process | 1.0 | 755 | 180 | – | 29 | 35 | 999 | 244 |
| TOTAL: | | 1126 | 5329 | 3565 | 80 | 110 | 6645 | 5519 |
| *Scheme 3*: | | | | | | | | |
| electric smelting of scrap | 1.0 | 2127 | 233 | 233 | 690 | 35 | 3085 | 959 |

**Table 8.2** Change in the average consumption of energy resources for the ferrous metallurgy industry as a function of the ratio of the production scheme outputs

| Index | Contribution of technological schemes | | Energy [kWh/t of steel] | | | | | | |
|---|---|---|---|---|---|---|---|---|---|
| | Level I | Level II | Chemical energy of raw material | Chemical energy of fuel: Total | Chemical energy of fuel: Including coke, electrodes | Electric | Heat | Total consumption: Total | Total consumption: Including conventional forms |
| *According to technological scheme (level I)* | | | | | | | | | |
| 1 | 0.46 | 0.24 | 1307 | 5394 | 2690 | 51 | 91 | 6843 | 5536 |
| 2 | 0.36 | 0.40 | 1126 | 5329 | 3565 | 80 | 110 | 6645 | 5519 |
| 3 | 0.18 | 0.36 | 2127 | 233 | 233 | 690 | 35 | 3085 | 958 |
| *Average for industry:* | | | | | | | | | |
| level I | | | 1391 | 4430.4 | 2543.4 | 177.7 | 87.7 | 6086.8 | 4695.8 |
| level II | | | 1532 | 3496.5 | 2139.5 | 294.1 | 78.3 | 5400.9 | 3868.9 |
| *Change:* | | | | | | | | | |
| kWh | | | +141 | −933.9 | −403.9 | +116.4 | −9.4 | −685.9 | −826.9 |
| kg c.e. | | | – | −114.7 | −49.6 | +37.2 | −1.4 | – | −78.9 |

ton of smelted steel, equivalent to 37.2 kg c.e. Therefore, the net savings of energy resources amount to 79.0 kg c.e. per ton of smelted steel.

Note that a change in the steelmaking industry will also affect related industries, in particular, coke and coking by-products production. A decrease in coke consumption of 50 kg per ton of steel will result in the reduction of energy consumption for coke production, later on this should be taken into consideration.

### 8.1.2 Impact of Improvements in Current Production Processes

In this section we report the results of calculations of the influence of improvements in the individual production stages of a given technological scheme for metallurgical production of liquid steel. It was assumed that in the future large-scale energy conservation measures would be successfully implemented. This allowed us to lower the current (base) level of energy consumption. The specific improvements are given below.

**Agglomeration.** Utilization of the sensible heat of the sinter with heated air that is fed into the hearth of sintering machines at a temperature of 773 – 973 K allows the decrease of specific fuel consumption by 5 kg c.e. [18, 41, 168]. In the calculations the volume of sinter production for the considered period is taken to be equal to 30 million t.

**Blast-furnace process.** a) Elevation of blast-furnace air temperature. According to [18, 41, 168] each 10-degree increase in the blast air temperature decreases the coke consumption by 0.2 %. We assumed an elevation of the average blast temperature for the considered period to be not less than 50 degrees.

b) Decrease in the blast-furnace gas losses through leaks. Furnaces which reduce losses by a factor of 2 are already available.

c) Introduction of systems of using the excess pressure and sensible heat of blast-furnace gas. Installation of gas heat-recovery turbines without compressors makes it possible to produce 40 – 60 kWh of electric power per ton of iron. The calculations show that this measure allows the production of an additional 10 – 12 kWh of electric power per ton of iron, for the considered period.

d) Introduction of evaporative cooling of blast furnaces. These systems potentially enable the production of 70 – 75 kWh of heat energy per ton of iron. It is conservatively assumed that in the considered period evaporative cooling will be introduced in a number of blast furnaces such that 10 – 15 kWh of heat energy per ton of iron on average for the industry can be produced.

e) Increase in the fraction of dry quenching done in the blast furnaces. According to [41, 168] the use of dry quenching in the blast furnaces improves their output and decreases the coke consumption per ton of iron by 2 – 3 %.

**Open-hearth process.** a) Intensification of the process by feeding oxygen and elevation of the blast air temperature. This results in the reduction of the smelting period and increased output that provides the saving of fuel consumption rate of not less than 5 %.

b) Increase in the extent of thermal secondary energy resource utilization. Increase in efficiency of already operating recovery equipment and installation of the new apparatus allow an additional 90 kWh of heat energy per ton of steel to be produced.

**Oxygen-converter process.** The trapping of converter gases. This enables the production of 17 – 20 kg c.e. of gaseous fuel with $i_{ch}$ = 2200 – 2300 kcal/m$^3$ per ton of converter steel. For the considered period it is assumed that the number of such installations introduced on the converters will yield a volume of 30 million t of smelted steel.

**Electric smelting.** Preparation and heating of scrap loaded into electric furnaces shorten the smelting duration, increase the output and decrease the consumption of electric power by 12 – 15 %, depending on the furnace capacities. It is assumed in the calculations that these improvements will decrease the consumption of electric power per ton of steel by 10 %.

The implementation of the enumerated measures in the indicated volumes leads to a considerable decrease in the specific energy consumption. Table 8.3 shows the summary decrease of the average specific consumption of energy resources due to both a restructuring of the sector (Sect. 8.1.1) and the improvement of existing processes and equipment.

## 8.2 Coke and Coking By-product Production

In this branch of metallurgy, at least three ways to decrease the energy consumption are

a) improvement of the equipment for heating and drying the furnace charges;
b) improvement of the furnaces of the coke-oven batteries;
c) replacing wet quenching with dry quenching.

According to [41, 168] implementation of point a) (in particular, installing fluidized bed furnaces) could save energy between 4 and 12 kg c.e. per ton of coke produced.

The furnaces of currently used coke-oven batteries are usually equipped with recuperators after which the temperature of waste gases reaches 575 – 625 K. An increase in the degree of heat recuperation in these furnaces in coke-oven batteries with and without recuperators could decrease the fuel consumption for heating coke-oven batteries by 5 %.

Point c), introduction of dry quenching, could be potentially a more effective energy conservation measure than the first two alternatives. It enables utilization of coke heat from the coke-oven batteries to generate 0.4 – 0.45 t of steam at a pressure of 4.0 MPa and a temperature of 723 K (0.3 – 0.35 Gcal) per ton of coke.

**Table 8.3** Change in the average consumption of energy resources for steel production

| Index | Contribution of technological schemes | | Energy [kWh/t of steel] | | | | | | |
|---|---|---|---|---|---|---|---|---|---|
| | | | Chemical energy of raw material | Chemical energy of fuel | | Electric | Heat | Total consumption | |
| | Level I | Level II | | Total | Including coke, electrodes | | | Total | Including conventional forms |
| *According to technological scheme (level I)* | | | | | | | | | |
| 1 | 0.46 | 0.24 | 1307 | 5186 | 2646 | 45 | – | 6538 | 5231 |
| 2 | 0.36 | 0.40 | 1126 | 5017 | 3505 | 72 | 103 | 6318 | 5192 |
| 3 | 0.18 | 0.36 | 2127 | 233 | 233 | 620 | 35 | 3015 | 888 |
| *Average for industry:* | | | | | | | | | |
| level I | | | 1391 | 4430.4 | 2543.4 | 177.7 | 87.7 | 6086.8 | 4695.8 |
| level II | | | 1532 | 3319 | 2090 | 265 | 53 | 5169 | 3637 |
| *Change:* | | | | | | | | | |
| kWh | | | +141 | –1111.4 | –453.4 | +87.3 | –34.7 | –917.8 | –1058.8 |
| kg c.e. | | | – | –136.6 | –55.7 | +27.9 | –5.2 | – | –113.8 |

In our calculations of the potential amounts of energy conserved, we have assumed that in the future, there will be twice as many dry quenching installations as there are now (i.e., in 1985). The results of these calculations predict

- the total fuel consumption per ton of coke decreases by 5 %;
- the consumption of heat energy drops by 175 kWh.

The detailed results, sorted according types of energy are given in Table 8.4.

**Table 8.4** Change in consumption of energy resources in coke and coking by-product production [kWh/t of coke]

| Energy resources | Level I | Level II | Change [kWh] | Change [kg c.e.] |
|---|---|---|---|---|
| Chemical energy of raw material | 15208 | 15208 | | |
| Chemical energy of fuel | 1000 | 950 | −50 | −6.1 |
| Electric energy | 32 | 32 | – | – |
| Heat energy | 240 | 65 | −175 | −26.3 |
| Total energy, | 16480 | 16255 | −225 | – |
| including conventional forms | 1272 | 1047 | −175 | −32.4 |

## 8.3 Rolled Stock

The method of continuous steel casting, developed in the USSR, is one of the primary means of improving the rolling process. Introduction of continuous casting yields

- considerable decrease in metal waste in rolled stock production (by 30 – 40 %);
- elimination of break-down mills and pit furnaces in the rolled stock shops, shops of preparing ingot moulds and stripping in the steelmaking shops;
- reduction of fuel and electricity consumption due to elimination of one process stage (roll cogging) in rolling (35 – 40 kg c.e. and 20 kWh per ton of steel);
- decrease in the maximum electric load of the plant by 3 – 4 %;
- increase in the productivity of labour in steelmaking and rolling complex by 20 – 25 %;
- decrease in the capital investment in rolling by 5 – 7 %.

In our calculations, we used available data to set the volume of steel produced by continuous casting at 19 – 20 % of all steel produced in the USSR in 1985. For the future level of continuous casting use we chose 50 %. This also increases output of sound goods of metal from 0.71 to 0.74, that is, 60 kg of liquid steel less is needed to produce 1 ton of rolled stock.

Other possibilities for reducing fuel noncontinuous casting facilities primarily involve the temperature elevation of the metal loaded into the furnaces and improvement of the furnace efficiency. According to [41, 168] raising the metal temperature by 100°C saves 7 – 8 kg c.e. of coal equivalent per ton of metal. One method is to place the ingots with the nonsolidifying part into soaking pits, whose temperature is somewhat higher than the temperature of the metal solidifying and to hold them there, until they acquire the temperature needed for rolling [41, 168]. In this case the fuel consumption decreases by 40 – 45 %. Moreover, the period from the beginning of casting to loading into the pits is shortened from 132 – 145 min to 55 – 95 min, the service life of ingot moulds and the turnover rate of the rail transport increase by 10 – 15 %.

One of the methods for improving the efficiency of heating furnaces and decreasing their fuel consumption is the recuperation of heat in waste gases. By [41, 168] raising the temperature of the air that is fed into the furnaces by 100°C results in a fuel saving of 3.5 to 5 kg c.e. per ton of the heated metal.

A way to reduce the heat energy consumption in rolling is to increase the level of secondary energy resources utilization, via improvement of the efficiency of available heat recovery equipment as well as installation of new devices.

In our calculations of the energy resource savings in rolling steel we have obtained the following results:

- introduction of continuous casting machines reduces fuel consumption from 125 to 108 kg c.e. and the electric power consumption from 105 to 99 kWh per ton of metal;
- increase in the temperature of the metal loaded into the heating furnaces and improvement of furnace efficiency reduces fuel consumption by 13 kg c.e.;
- improvement of the efficiency of available heat recovery equipment and installation of new facilities allow additional production of about 410 kWh of heat per ton of metal and hence, the complete meeting of heat demands of rolling and even some reduction of its consumption by other process stages.

The average material and energy balances for rolling were calculated, again using 1985 as the base level, and the specific consumption of energy resources was determined. Their results are shown in Table 8.5.

**Table 8.5** Change in consumption of energy resources in rolling [kWh/t of rolled stock]

| Energy resources | Level I | Level II | Change [kWh] | Change [kg c.e.] |
|---|---|---|---|---|
| Chemical energy of fuel | 1020 | 775 | −245 | −30.1 |
| Electric energy | 105 | 99 | −6 | −1.9 |
| Heat energy | 75 | – | −75 | −11.3 |
| Total | 1200 | 874 | −326 | −43.3 |
| including conventional forms | 1200 | 874 | −326 | −43.3 |

**Table 8.6** Change in the structure of consumed energy resources and of energy saved at the future level (per ton of rolles steel) [kWh]

| Index | Chemical energy of raw material | Chemical energy of fuel | | Electric energy | Heat energy | Total consumption | |
|---|---|---|---|---|---|---|---|
| | | Total | Including coke, electrodes | | | Total | Including conventional forms |
| Level I, in steel production | 1961.3 | 6246.9 | 3586.2 | 250.6 | 123.7 | 8582.5 | 6621.2 |
| coke and coking by-products | 6752.4 | 444.0 | – | 14.2 | 106.7 | 7317.2 | 564.8 |
| rolling | – | 1020.0 | – | 105.0 | 75.0 | 1200.0 | 1200.0 |
| TOTAL: | 8713.7 | 7710.9 | 3586.2 | 369.8 | 305.3 | 17099.7 | 8386.0 |
| Level II, in steel production | 2068.2 | 4480.6 | 2821.5 | 357.8 | 71.6 | 6978.2 | 4910.0 |
| coke and coking by-products | 5444.5 | 340.1 | – | 11.5 | 23.3 | 5819.4 | 374.9 |
| rolling | – | 775.0 | – | 99.0 | – | 874.0 | 874.0 |
| TOTAL: | 7512.7 | 5595.7 | 2821.5 | 468.3 | 94.9 | 13671.6 | 6158.9 |
| Total reserve of energy saved (–) overconsumed (+) kWh | –1201.0 | –2115.2 | –764.7 | +98.5 | –210.4 | –3428.1 | –2227.1 |
| kg c.e. | – | –259.9 | –93.9 | +31.5 | –31.7 | – | –260.1 |

The results of Tables 8.3–5 are summarized and compared again in Table 8.6. This table shows the total amount of possible energy resources which can be conserved per ton of rolled stock steel when all of the above factors are taken into account. It should be noted that the obtained figures represent the minimal estimate of the amount saved since

- improvements in extraction methods, primary ore processing and concentration were not taken into account;
- not all possible energy conservation measures were considered in the auxiliary production processes;
- the most pessimistic estimate of the amount of energy conserved was always taken;
- methods which involve only a change in the type of energy carrier and yield no net energy conservation were not included.

Thus, a more complete inclusion of energy conservation mesures should lead to an even higher estimate of the amount of energy that could potentially be saved.

## 8.4 Influence of Other Parameters

As was shown above, numerous factors influence the energy consumption, process temperature and pressure, quality of raw material, structure of the technological schemes, extent of introducing advanced technologies (continuous steel casting, dry quenching of coke, production of pellets, etc.) and many others.

The previous sections considered the estimation of the amount of energy saved in the production of steel, rolled stock, and coke due to measures whose impact on the energy consumption can be determined on the basis of the complete energy balances.

There is also a large category of averaged parameters, whose effect can be determined without the use of energy balances. Among them are the mean content of iron in ore (which depends on the probability of discovery of new sources and their preparation to operation), the content of iron in concentrate (which depends on technologica progress in enrichment processes), and the structure of the technological schemes.

A simple algorithm has been written to estimate the effect of these factors on energy consumption. It enables the evaluation of the energy consumption for the primary products, in the case considered here, these are steel in rolled stock, foundry pig iron and coke. The algorithm includes all production stages from natural raw material production through to the final product.

The algorithm allows the calculation of a large number of variants with the variable parameters changing over wide ranges. Thus, the most probable zone of change in energy consumption can be found, which is then narrowed by more detailed calculations with smaller adjustments in the parameters.

Although the qualitative effects of almost all factors are known, the problem is to find the quantitative estimation of the amount of energy consumed, or conserved. It is necessary to determine both the effect of a parameter change on the energy consumption for a given individual process and on the whole technological scheme. In each calculation, only a single parameter, for example, the content of iron is ore, ratio of agglomerate and pellets in the charge of the smeltery blast furnace, was varied and the others were held fixed.

The procduction of useful metal, whose volume is the inverse of the liquid steel consumption per ton of the finished rolled stock, is one of the most important indicators of the efficiency of the rolling process (the process also depends on the method of steel casting). Both the fraction of continuous casting and the production of useful metal were varied in the calculations. In the best variant the liquid steel consumption per ton of rolled stock was equal to 1.15 t, this corresponds to the current record level.

The maximum fraction of steel poured by continuous casting was taken to be 0.8. However, this value was corrected and coordinated with the fraction used in open-hearth processing depending on the smelting method. The sum of the two fractions was set to 1, since continuous casting is incompatible with open hearths.

Other improvements are also included in the model. Four different methods of ore concentration with different coefficients of metal extraction into the concentrate are compared. Three methods for producing pellets of different types are also taken into account roasted, fluxed on carbonate and on cement binder. However, there are no sufficiently reliable data on many new processes, therefore, their impact on the future energy consumption of ferrous metallurgy cannot be evaluated.

Numerous calculations, with varying initial parameters allowed some quantitative estimates to be obtained. First of all, the total energy consumption in steel production rolled stock was evaluated. The fraction of steel that is poured by continuous casting machines ($\beta_{\text{CCM}}$) and the level of efficiency of the machines of providing different outputs of sound goods ($\gamma$), as well as the fraction of pellets in the charge of the blast-furnace ($\beta_{\text{pel}}$) have the greatest influence on the value of the energy consumption. Figure 8.1 shows the results. Three technological schemes of rolled stock production were compared, open-hearth, oxygen-converter, and electric scrap smelting. The fraction of open-hearth steel $\beta_{\text{oh}} = 0.24$; oxygen-converter steel $\beta_{\text{oc}} = 0.40$; electrically smelted scrap steel $\beta_{\text{es}}^{\text{scr}} = 0.36$.

As was shown above, forecasting the structure of steel production technologies is an extraordinarily complex problem. Therefore, strongly differing opinions on the degree to which energy can be conserved are found in the literature. We are aware of models in which the fractions of each of the steel production methods vary by factors of two or even three. Some are more optimistic and others less optimistic about the value of substituting open-hearth processes by oxygen-converter and electric steel smelting.

For these reasons we have carried out our calculations so that the parameters range over very wide limits. The fraction of steel poured in continuous casting

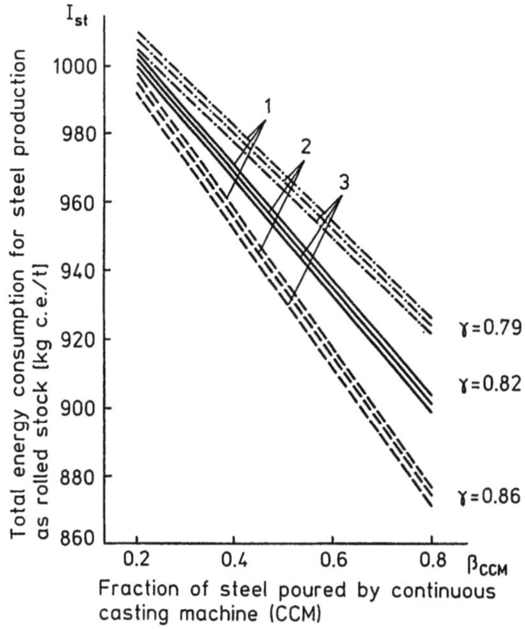


**Fig. 8.1.** Influence of the increase in continuous casting output on the total energy consumption in steel production as rolled stock. Fraction of pellets in the furnace charge curve *1* – 0.3; *2* – 0.4; *3* – 0.8

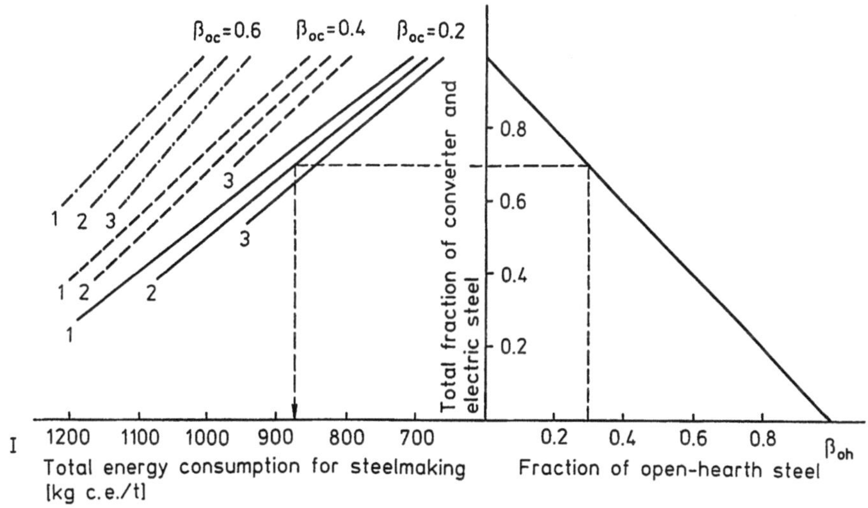


**Fig. 8.2.** Estimation of the total specific energy consumption in rolled stock steel production versus the structure of the main steelmaking processes. The fractions of steel poured by continuous casting are *1* – 0.2; *2* – 0.4; *3* – 0.8

was varied in addition to the technological structure. The fraction of electric steel production using iron-rich pellets was not varied, since this technology is used to smelt high quality steels and will not compete with the main processing methods.

The obtained results allow a diagram such as shown in Fig. 8.2 to be constructed. This graph can be used to make tentative estimations. Under the conditions of minimum energy consumption and the most efficient equipment i.e., with the maximum possible fraction of continuous casting $\beta_{CCM} = 0.8$, the output of sound goods of metal $\gamma = 0.86$, and the fraction of pellets in the blast-furnace charge of $\beta_{pel} = 0.8$. The energy consumption is ~850 kg c.e. per ton of steel.

Similar studies were carried out for cast iron and coke. In the first case the content of Fe in the ore and the fraction of pellets in the furnace charge were varied. In the second case the fraction of dry coke quenching was varied so that the production of a significant amount of heat energy (approximately 300 – 350 kWh per ton of steel) was observed.

Thus, this technique allows rather quick evaluations of energy consumption for the primary products. However, the assumptions simplify the problem to the point that linear behaviour, not observed in reality, is seen. Only the use of complete energy balances will account for non-linearity. This ist the challenge for future work.

# References


1. J.E. Ahern: *The Exergy Method of Energy Systems Analysis* (John Wiley, New York 1980)
2. J. Ahrendts: Reference States, Energy, **5**(5), 667–677 (1980)
3. J. Ahrendts: *The Exergy of Chemically Reacting Systems* (in German), VDI-Forschungsheft 597 (VDI, Düsseldorf, 1977)
4. *Analysis of Production and Consumption Structure of Ferrous Metals in the USSR* Ed. R.G. Kamalov (in Russian) (Metallurgiya, Moscow 1980)
5. S.M. Andoniev: *Evaporative Cooling of Metallurgical Furnaces* (in Russian) (Metallurgiya, Moscow 1970)
6. V.P. Andreev, V.I. Rushchuk: "Energy-Saving Technological Processes in Non-Ferrous Metallurgy" (in Russian) Promyshlennaya energetica, **40**(3), 2–3 (1983)
7. L.P. Andreev, G.N. Kostenko: "Efficiency and Degree of Thermodynamic Perfection of Heat-Exchange Units" (in Russian) Izv. Vuzov. Energetika, **1**(5), 148–150 (1957)
8. A.I. Andryushchenko: *The Ability to Perform Work in Thermodynamic Systems* (in Russian), (Saratovskyi Avtomobilno-Dorozhnyi Institut, Saratov, 1956)
9. A.I. Andryushchenko: *The Principles of Technical Thermodynamics of Real Processes* (in Russian) (Vysshaya Shkola, Moskow 1975)
10. A.I. Andryushchenko: *Thermodynamic Calculations of Optimal Parameters of Thermal Power Plants* (in Russian) (Vysshaya Shkola, Moskow 1963)
11. H.D. Baehr: "Definition and Calculation of Exergy and Anergy" (in German), Brennst. Wärme Kraft, **17**(1), 1–6 (1965)
12. H.D. Baehr: "Exergy – A Useful Tool for Chemical Engineers?" Proc. 2nd World Congr. Chem. Eng., Montreal, Vol. 2, 238–240 (1981)
13. H.D. Baehr: *Thermodynamics* (in German), 4th ed., (Springer Berlin, 1978)
14. H.D. Baehr, E.F. Schmidt: "Definition and Calculation of Fuel Exergy" (in German), Brennst. Wärme Kraft, **15**(8), 375–381 (1963)
15. O.N. Bagrov, B.M. Kleshko, V.V. Mikhailov: *Energy Aspects of the Main Processes in Non-Ferrous Metallurgy* (in Russian) (Metallurgiya, Moscow 1979)
16. Yu.V. Baimakov, A.N. Zhurin: *Electrolysis in Hydrometallurgy* (in Russian) (Metallurgiya, Moscow 1963)
17. Yu.M. Bakhramov: *Economic Efficiency of Using Wastes in Steel Production* (in Russian) (Leningrad University Press 1982)
18. N.P. Banny, A.A. Fedorov, P.A. Shiryaev, L.N. Roitburd: *Economic Aspects of Ferrous Metallurgy* (in Russian) (Metallurgiya, Moscow 1978)
19. R.E. Bellman: *Dynamic Programming* (in Russian) (Inostrannaya Literatura, Moscow 1960)
20. R.E. Bellman, S.E. Dreyfus: *Applied Dynamic Programming* (in Russian) (Nauka, Moscow 1963)
21. A.I. Belyaev: *Metallurgy of Light Metals* (in Russian) (Metallurgiya, Moscow 1970)
22. T.T. Berdysheva: "Production and Use of Carbonate-Base Pellets in the USA (in Russian) Chermetinformatsiya, ser. 3, vyp. 3, (1969)

23. A.I. Berezhynski, I.M. Tinyakova: "The Problem of Additional Heating of Converter Waste-Heat Boilers" (in Russian), Promyshlennaya Energetika, **38**(5), 40–42 (1981)
24. Z.L. Berlin: *Rational Use of Secondary Energy Resources in Non-Ferrous Metallurgy* (in Russian) (Metallurgiya, Moscow 1972)
25. Yu.I. Blank, I.V. Onufriev: "Chemical Exergy of Substances, Solutions, Electrolytes and Ions" (in Russian), Promyshlennaya Elektrotekhnika, No. 6, 69–73 (1987)
26. A.V. Borodulin, I.I. Kobeza, A.F. Kovtun, V.S. Stepanov: "Environmental Aspects of Metallurgical Production" (in Russian), Metallurgicheskaya i gornorudnaya Promyshlennost (Dnepropetrovsk, izd-vo "Promin") No. 1 (155), 55–57 (1990)
27. F. Bošnjaković: *Technical Thermodynaics* (in Russian), Vol. 1–2 (Gosenergoizdat, Moscow-Leningrad 1956)
28. F. Bošnjaković: "Reference Levels of Exergy of Chemically Reacting Systems" (in German), Vorsch. Ingenieurw. **29**(5), 151–152 (1963)
29. I. Boustead, G.F. Hancock: *Handbook of Industrial Energy Analysis* (Ellis Harwood, London, 1979)
30. D. Boyadjyiev: "Exergy Theory and Possibilities of Application" (in Bulgarian), Energetica, **23**(4), 8–12 (1972)
31. M. Brauer, R. Jeschar: "The Significance of Exergy for Thermodynamical Investigations of the Blast-Furnace Process" (in German), Arch. Eisenhüttenw. **34**(1), 9–16 (1963)
32. E.V. Britske, A.F. Kapustinski, B.K. Veselovski, L.M. Shamovski, L.G. Chentsova, B.I. Anvaer: *Thermodynamic Constants of Non-Organic Substances* (in Russian) (Izd-vo AN SSSR, Moscow 1949)
33. V.M. Brodyanski: *Energy and Economy of the Complex Fractionation of Air* (in Russian) (Metallurgiya, Moscow 1965)
34. V.M. Brodyanski: *Exergy Method of Thermodynamic Analysis* (in Russian) (Energiya, Moscow 1973)
35. V.M. Brodyanski, J.L. Leites, Y.G. Karpova: "Selection of Reference Levels for Exergetic Analysis of Chemical Processes" (in Russian), Teor. Osn. Khim. Tekhnol. **5**(6), 858–862 (1971)
36. V.M. Brodyanski, M.V. Sorin: "Models of Environment for Chemical Exergy Calculation" (in Russian), Teor. Osn. Khim. Tekhnol. **18**(6), 816–824 (1984)
37. D. Buimowici: "Maximum Work and Energy of Fuels" (in German), Allg. Wärmetech. **8**(9), 189–194 (1958)
38. J.A. Campbell: *Modern General Chemistry* (in Russian) (Mir, Moscow 1975)
39. P.F. Chapman: "Energy Content and Availability of Metals", IEE Proc. **130**(6), Pt. A, 354–355 (1983)
40. P.F. Champman: "The Energy Costs of Producing Copper and Aluminium from Primary Sources", Metals and Mater. **8**(2), 1974)
41. M.H. Chiogioji: *Industrial Energy Conservation* (Marcel Dekker, New York, Basel 1979)
42. V.G. Chernyavski, V.O. Mednik, E.S. Durov: "Reserves for Increasing the utilization of Thermal Secondary Energy Resources Use at Metallurgical Plants in the Ukraine (in Russian) Promyshlennaya Energetika, **39**(2), 5–7 (1982)
43. *Concise Reference-Book of Physico-Chemical Values* (in Russian) (Khimiya, Leningrad 1972)
44. D.A. Diomidovski: "The Theory of Furnaces in Non-Ferrous Metallurgy" (in Russian), Tsvetnye Metally, **31**(10), 24–26 (1956)
45. D.A. Diomidovski, L.M. Shalygin, A.A. Galnbek, I.A. Yuzhaninov: "Calculations of Pyroprocesses and Furnaces in Non-Ferrous Metallurgy" (in Russian), (NTIL po tsvetnoy i chernoy metallurgii, 1963)
46. S.B. Dudenkov: "Methods for Improving the Filtration Processes in Non-Ferrous Metallurgy" (in Russian), Tsvetnye Metally, **49**(7) (1974)

47. G.E. Edelgauz: *Accuracy, Reliability and Stability of Indices* (in Russian), (Leningrad State University 1971)
48. K. Smith (ed.): *Efficient Use of Electric Energy* (in Russian), (Energoizdat, Moscow 1981)
49. A.P. Egorichev, B.V. Sazanov, L.N. Albul: "Output Regulation of Waste-Heat Boilers" (in Russian), Promyshlennaya Energetika, **41**(4), 9–11 (1984)
50. G. Eger (ed.): *Electrical Metallurgy of Water Solutions (in Russian), (Metallurgiya, Moscow 1966)*
51. N. Elsner, W. Fratzscher: "On the Problem of Irreversibilities in Thermal Processes" (in German), Wiss. Z. Hochsch., Verkehrsw. Dresden, **5**(1), 33–47 (1957); Energietechnik, **8**(8) 345–349 (1958)
52. N. Elsner, W. Fratzscher: "Role of exergy analysis in studying thermal power installations on the examples of a waste-heat boiler, thermal power plant and steam engine (in German), Wiss. Z. Hochsch. Verkehrswes., Dresden, **5**(3), 437–446 (1957)
53. V.M. Brodyanski (ed.): *Energy and Exergy* (in Russian) (Mir, Moscow 1968)
54. N.G. Zalogin (ed.): *Energy and Environmental Protection* (in Russian) (Energiya, Moscow 1979)
55. R.B. Evans: *A Proof that Exergy is the Only Consistent Measure of Potential Work*, Ph. D. Thesis, Thayer School of Engineering, Dartmouth College, Hannover, N.H. (1969)
56. R.B. Evans: "Thermoeconomic Isolation and Exergy Analysis", Energy, **5**(8/9), 805–821 (1980)
57. *Exergy Method and its Applications* Ed. V.M. Brodyanski (in Russian) (Mir, Moscow 1967)
58. M.A. Eydenson: *Metallurgy of Magnesium and Other Light Metals* (in Russian) (Metallurgiya, Moscow 1964)
59. L.T. Fan, J.H. Shieh: "Multiobjective Optimal Synthesis", in *Efficiency and Casting: Second Law Analysis Processes* (American Chemical Society, Washington, D.C., 1983) pp. 307–347
60. N.P. Fedotiev, A.F. Alabyshev, A.P. Rotinyan, P.M. Vyacheslavov, P.B. Zhivotinski, A.A. Galnbek: *Applied Electrochemistry* (in Russian) (Khimiya, Leningrad 1967)
61. W. Fratzscher: "Driving Force and Irreversibility in Chemical Engineering" (in German), Wiss. Z. Tech. Hochsch. Leuna-Merseburg, **19**(4), 557–562 (1977)
62. W. Fratzscher, K. Michalek: "Evaluation of the Waste Energy Balances of a Large Chemical Plant" (in German), Energieanwendung, **31**(6), 202–207 (1982)
63. W. Fratzscher, D. Schmidt: "On the Determination of Maximum Work of Flame Reactions" (in German), Wiss. Z. Th., Dresden, **10**(1), 183–191 (1961)
64. W.L.R. Gallo, L.F. Milanez: "Choice of a Reference State for Exergetic Analysis", Energy, **15**(2), 113–121 (1990)
65. R.M. Garrels, C.L. Christ: *Solutions, Minerals and Equilibria* (in Russian), (Mir, Moscow 1986)
66. L.M. Gazaryan: *Pyrometallurgy of Copper* (in Russian) (Metallurgiya, Moscow 1965)
67. J.W. Gibbs: *Thermodynamic Works* (in Russian) (Gostekhteorizdat, Moscow 1950)
68. J.W. Gibbs: "On the Equilibrium of Heterogeneous Substances", in *Scientific Papers*, Vol. 1 (Dover, New York, 1961)
69. D.P. Gokhshteyn: *Entropy Method of Calculation of Energy Losses* 2nd. Ed. (in Russian) (Izd-vo Gosenergo, Moscow-Leningrad 1963)
70. G. Gouy: "About Available energy" (in French), J. Phys. II, **8**, 501–518 (1889)
71. P. Grassmann: "Energy and Exergy: The Expression of Losses by Means of Exergy Balances" (in German), Verfahrenstechnik, **13**, 28–31 (1979)
72. P. Grassmann: "On the General Definition of Efficiency" (in German), Chem. Ing. Tech. **22**(4), 77–80 (1950)

73. L.D. Gritsuk: "Reserves and Methods for Increasing the Use of Secondary Energy Resources" in *Efficiency Increase of Thermal Insulating Equipment and Cooling Systems in Ferrous Metallurgy (in Russian) (Metallurgiya, Moscow 1982) pp.55–57.*
74. R.W. Haybood: *Analysis of Engineering Cycles* (in Russian) (Energiya, Moscow 1979)
75. S. Ikumi, C.D. Luo, C.Y. Wen: "A Method of Estimating Entropies of Coals and Coal Liquids", Can. J. Chem. Eng. **60**, 551–555 (1982)
76. M. Ishida: "Hierarchical Structure of Process System", in *Efficiency and Casting: Second Law Analysis of Processes* (American Chemical Society, Washington, D.C. 1983) pp. 173–211
77. M. Ishida, K. Kawamura: "Energy and Exergy Analysis of a Chemical Process with Distributed Parameters Based on the Enthalpy-Direction Factor Diagram", Ind. Eng. Chem. Process Des. Dev. **21**(4), 690–695 (1982)
78. M. Ishida, N. Nakagawa: "Exergy Analysis of a Pervaporation System and its Combination with a Distillation Column Based on an Energy Utilization Diagram". J. Membrane Sci. **24**, 271–283 (1985)
79. L.B. Ivanov, G.A. Kudinov, E.E. Lysenko, V.I. Trusov, V.N. Kopylov: "Prospects of Using Low-Pressure Heat in a Power Loop with a Working Medium with a Low Boiling Point", in *Efficiency Increase of Thermal Insulating Equipment and Cooling Systems in Ferrous Metallurgy* (in Russian) (Metallurgiya, Moscow 1982)
80. R. Jeschar: "Exergy Balance of Tunnel Furnaces" (in German), Ber. Deut. Keram. Ges. e.V. **40**(11), 596–602 (1963)
81. R. Jeschar; R. Görgen: "Contribution to the Use of Exergy Balances at Metallurgical Plants", Stahl-Eisen, No. 12 (1965)
82. J.I. Kalynina, V.M. Brodyanskyi: "Thermoeconomic Method of Costs Partition in Multipurpose Technical Systems" (in Russian), Izv. Vuzov. Energetika **17**(3), 58–63 (1974)
83. H. Kameyama, K. Yoshida: "On the Concept of Standard Exergy", Energy Dev. Jpn. (Chicago) **3**(1) 41–49 (1980)
84. H. Kameyama, K. Yoshida, S. Yamauchi, K. Fueki: "Evaluation of Reference Exergies for the Elements", Applied Energy, **11**(1), 69–83 (1982)
85. M.Kh. Karapetiyants, M.L. Karapetiyants: *The Main Thermodynamic Constants of Non-Organic and Organic Substances* (in Russian) (Khimiya, Moscow 1968)
86. A.A. Kazakov: *Continuous Steelmaking Processes* (in Russian) (Metallurgiya, Moscow 1977)
87. G.A. Khachkuruzov: *Fundamentals of General Chemical Thermodynamics* (in Russian) (Vysshaya Skola, Moscow 1979)
88. M.V. Kirpichev: "Energy Balance of Thermal Installations" (in Russian) Izv. AN SSSR, OTP, No. 12 (1949)
89. J.H. Keenan: "Availability and Irreversibility in Thermodynamics", Br. J. Appl. Phys. **2**, 183–192 (1951)
90. J.H. Keenan: *Thermodynamics* (in Russian) (Gosenergoizdat, Moscow 1963)
91. W.K. Kenney: "Second Law Analysis to Improve Industrial Processes", in *Efficiency and Casting: Second Law Analysis of Processes* (American Chemical Society, Washington, D.C. 1983) pp. 51–67
92. V.A. Kirillin, V.V. Sychev, V.A. Sheindlin: *Technical Thermodynamics* (in Russian) (Nauka, Moscow 1979)
93. V.F. Knyazev: "Direct Production of Iron and Steel", in *Results of Science and Engineering, Ser. Iron and Steel Production*, Vol. 12 (Izd-vo VINITI, Moscow 1980)
94. S.M. Komkov: "Use of Heat Recovery Installations as Gas Cleaners for Reducing Emissions into Air and Fuel Conservation", Rep. at EEK OON Symp. on Gas Industry and Environment, Minsk (June, 1977)
95. A.F. Kotin: "On the Physical Essence of Exergy", Izv. Vuzov, Energetika, **9**(7), 49–57 (1965)

96. I.Yu. Kozhevnikov: *Coke-Free Metallurgy of Iron* (in Russian) (Metallurgiya, Moscow 1970)
97. V.V. Krivoruchenko, M.A. Korobov: *Heat and Energy Balances of Electrolyzers* (in Russian) (Metallurgizdat, Moscow 1963)
98. N.F. Krupkin, A.M. Leonova: "Autogeneous Processes and Energy Resource Conservation in Non-Ferrous Metallurgy" (in Russian), Tsvetnye Metally, **53**(2), (1978)
99. G.M. Krzhyzhanovski, B.I. Veits, E.A. Russakovski: "Fuel-Energy Balance" (in Russian), Vestnik Statistiki, No. 7 (1932)
100. O. Kubashewski, E.L. Evans: *Metallurgical Thermochemistry* (in German) (Verlag Technik, Berlin 1959)
101. O. Kubashewski, S.B. Olkok: *Metallurgical Thermochemistry* (in Russian), (Metallurgiya, Moscow 1982)
102. N.A. Kudryavaya, V.A. Novikov, M.I. Bystronogova: "Analysis of Operating Conditions of Dry-Quenching Machines of Coke Batteries", in *Use of Secondary Energy Resources and Cooling of Plants in Ferrous Metallurgy* (Metallurgiya, Moscow 1975) pp. 51–54
103. Yu.P. Kupryakov: *Autogeneous Smelting of Copper Concentrates in Flash Furnaces* (in Russian) (Metallurgiya, Moscow 1979)
104. B.N. Laskorin, B.V. Gromov, A.P. Tsygankov, V.N. Senin: *Waste-Free Technology in Industry* (in Russian), (Stroyizdat, Moscow 1986)
105. V.N. Leksin, N.V. Krupkin, L.G. Melnik: *Efficiency of Production Improvement in Non-Ferrous Metallurgy* (in Russian), (Metallurgiya, Moscow 1980)
106. V.N. Leksin, A.G. Tokareva: *Economic Aspects of the Use of Polymetallic Raw Materials* (in Russian), (Metallurgiya, Moscow 1968)
107. A.G. Lifshits: *Economic Efficiency of the Oxygen-Converter Production of Steel* (in Russian), (Metallurgiya, Moscow 1978)
108. F.I. Loskutov, A.A. Tseidler: *Calculations on Metallurgy of Heavy Non-Ferrous Metals* (in Russian), (Metallurgiya, Moscow 1963)
109. M.A. Lozano, A. Valero: "Calculating the Exergy of Industrial Solutions", Wint. Meeting ASME (Nov. 27–Dec. 2, 1988) pp. 45–51
110. M.A. Lozano, A. Valero: "Methodology for Calculating Exergy in Chemical Process", Wint. Meeting ASME (Nov. 27–Dec. 2, 1988) pp. 77–86
111. D. Merten: "Methods for Drawing Up Energy Balances and Their Applications", Mitt. Inst. Energ., No. 101, 747–758 (1968)
112. D.R. Morris, F.R. Steward: "Exergy Analysis of a Chemical Metallurgical Process", Met. Trans. **15B**, 645–654 (1984)
113. D.R. Morris, B.R. Amero, P.G. Evans, W. Petruk, D.R. Owens: "Reactions of Sinter in a Lead Blast Furnace", Met. Trans. B. **14**, 617–623 (1983)
114. D.R. Morris, F.R. Steward, P.G. Evans: "Energy Efficiency of a Lead Smelter", Energy, **8**(5), 337–349 (1983)
115. D.R. Morris, J. Szargut: "Standard Chemical Exergy of Some Elements and Compounds on the Planet Earth", Energy, **11**(8), 733–755 (1986)
116. R. Nitsch: "On the Theory of Cost Formation Using the Exergy Method", Sonderdruck Wiss. Ztschr. T.H.Chem., Leuna-Merseburg, 7(2), 100–104 (1965)
117. G.V. Nozdrenko, I.A. Kuzmina, N.D. Sycheva: "Distribution of Fuel Consumption During Heat and Electricity Production at a Cogeneration Plant" (in Russian), Izv. Sib. Otd. ANS SSR, Ser. Tekhn. Nauk, No. 8/2, 161–164 (1978)
118. V.D. Osipenko, V.P. Milevski, N.G. Shchekin: Utilization of Heat of the Waste Gases from Pit Furnaces, (in Russian), in *Efficiency Increase of Thermal Insulating Equipment and Cooling Systems in Ferrous Metallurgy* (Metallurgiya, Moscow 1982)
119. Ost-Rassow: *Manual of Chemical Technology* (in German) Vols. I, II (Johann Ambrosius Barth, Leipzig 1965)

120. I.Ch. Ozolyng, V.S. Stepanov: "Approximate Calculation Method of Chemical Energy and Exergy of Substances" (in Russian), Izv. Vuzov Energetika, **22**(10), 60–64 (1979)
121. I.Ch. Ozolyng, V.S. Stepanov, N.I. Tazhbekov: "Experiments in Drawing the Energy Balances of Energy Intensive Plants of Metallurgical and Chemical Industrie" (in Russian) in *Problems of Heat Engineering and Applied Thermal Physics*, Vol. 6 (Nauka, Alma-Ata 1970) pp. 78–88
122. I.Ch. Ozolyng, V.S. Stepanov, N.I. Tazhbekov: "Technique for Calculating the Energy Balances of Industrial Plants" (in Russian) in *Problems of Energy and Unified Energy Systems*, Vol. 3 (Nauka, Alma-Ata 1967) pp. 239–253
123. V.D. Pashkov, A.A. Zaretski, V.D. Teplov: "Ways for Improving the Schemes and Conditions of Work for Gas Turbines Using the Pressure of Blast-Furnace Gas" (in Russian), Promyshlennaya Energetika, **25**(10), 7–11 (1968)
124. V.S. Perepechko, N.A. Kudryavaya, N.A. Rogova, V.S. Menyakina: "Use of Secondary Energy Resources at Cherepovetski Metallurgical Plant" (in Russian), Promyshlennaya Energetika, **41**(2), 10 (1984)
125. V.M. Brodyanski (ed.): *Problems of Thermodynamic Analysis (Exergy Method)* (in Russian) (Mir, Moscow 1965)
126. G.V. Samsonov (ed.): *Properties of Elements* (in Russian) (Mir, Moscow 1976)
127. A.M. Pyatkin: "Methods of Economic Estimation and Choice of Efficient Energy Conservation Measures" (in Russian), Promyshlennaya Energetica, **43**(2), 4–7 (1986)
128. V.A. Rabinovich, Z.Ya. Khavin: *Concise Chemical Reference Book* (in Russian) (Khimiya, Leningrad 1977)
129. Z. Rant: "Exergy and Anergy" (in German), Wiss. Z. Tech. Univ. Dresden, **13**(4), 1145–1149 (1964)
130. Z. Rant: "Exergy, a New Word for 'Technical Available Work'" (in German), Forsch. Ing. Wes., **22**(1), 36–37 (1956)
131. Z. Rant: "Processes of Heating and the Second Law of Thermodynamics" (in Russian) in *Exergy Method and its Applications* (Mir, Moscow 1967) pp. 31–48
132. A. Rant: "Towards the Estimation of Specific Exergy of Fuels" (in German), Allg. Wärmetech., Vol. 10, N 9, pp. 172–176 (1961)
133. *Reference Book of Chemists*, Vol. I–III (in Russian) (Goskhimizdat, Moscow 1962)
134. *Reference Book on Heat Technology*, Vol. I (in Russian) (Energiya, Moscow 1975, 1976)
135. G.M. Reistad, R.A. Gaggioli: "Available Energy Costing", in *Thermodynamics: Second Law Analysis* (American Chemical Society, Washington, D.C., 1980)
136. L. Reikert: "Energy Transformation Through Chemical Processes" (in German), Chem. Eng. Sci., **47**, 48–51 (1975)
137. L. Reikert:: "The Efficiency of Energy Ultilization in Chemical Processes", Chem. Eng. Sci., **29**, 1613–1620 (1974)
138. H.H. Rosenbrock, C. Storey: *Computational Techniques for Chemical Engineers* (in Russian) (Mir, Moscow 1968)
139. A.E. Russakovski, A.I. Sheftel: *Energy in the National Economy* (in Russian) (GONTI, Moscow 1935)
140. J.D. Seader: "Thermodynamic Efficiency of Chemical Processes", in *Industrial Energy Conservation Manuals* (MIT Press, Cambridge, Mass. 1982)
141. N.A. Semenenko, L.I. Kuperman, S.A. Romanovski, N.M. Itskovich, L.N. Sidelkovski, L.K. Vukalovich: *Secondary Energy Resources and Energy-Technological Combining in Industry* (in Russian) (Vishcha Shkola, Kiev 1979)
142. V.V. Sergeev, N.V. Galitski, V.P. Kiselev: *Metallurgy of Titanium* (in Russian) (Metallurgiya, Moscow 1964)
143. J.H. Shieh, L.T. Fan: *Energy and Exergy Estimation Using the Group Contribution Method in Thermodynamics: Second Law Analysis* (American Chemical Society, Washington, D.C., 1983) pp. 351–371

144. J.H. Shieh, L.T. Fan: "Estimation of Energy (Enthalpy) and Exergy (Availability) Contents in Structurally Complicated Materials", Energy Sources, **6**(1–2), 1–46 (1982)
145. F.G. Shinskey: *Energy Conservation Through Control* (in Russian) (Mir, Moscow 1981)
146. A.P. Shitskova, Yu.V. Novikov, N.V. Klimkina, R.S. Gildenskiold, V.N. Shapritski: *Environmental Protection from Pollution by Ferrous Metallurgical Plants* (in Russian) (Metallurgiya, Moscow 1982)
147. L.N. Sidelkovski, E.Ya. Falkov: *Exergy Balances of Flame Processes* (in Russian) (Izd-vo MEI, Moscow 1965)
148. G.B. Sinyarev, N.A. Vatolin, B.G. Trusov, G.K. Moiseev: *Application of Computers for Thermodynamic Calculations of Metallurgical Processes* (in Russian) (Nauka, Moskow 1982)
149. V.I. Smirnov, A.A. Tseydler, I.F. Khudyakov, A.I. Tikhonov: *Metallurgy of Copper, Nickel and Cobalt*, Part I (in Russian) (Metallurgiya, Moscow 1964)
150. M.V. Sorin, V.M. Brodyanski: "Calculation of Chemical Exergy on the Basis of the Model of a Local Environment" (in Russian), Teor. Osn. Khim. Tekh. **19**(1), 91–99 (1985)
151. V.S. Stepanov: "Analysis of Energy Efficiency of Ferrous Metallurgy Processes and Methods for Improvements" (in Russian), Promyshlennaya Energetika, **36**(2), 9–12 (1979)
152. V.S. Stepanov: *Analysis of Energy Efficiency of Technological Processes* (in Russian) (Nauka, Novosibirsk 1984)
153. V.S. Stepanov: *Chemical Energy and Exergy of Substances* 2nd ed. (in Russian) (Nauka, Novosibirsk 1990)
154. V.S. Stepanov: "Concepts of Energy Conservation and Potential in Industrial Processes and Methods for Their Assessment" (in Russian), thirteenth Int. Conf. Ind. Energetics, Varna, Paper N A-2-27 (1990)
155. V.S. Stepanov: "Efficiency of Energy Use in Lead Production" (in Russian), Promyshlennaya Energetika, **41**(7), 36–38 (1984)
156. V.S. Stepanov: "Estimation of Energy Indices of Lead Production" (in Russian), Tsvetnye Metally, **56**(7), 9–12 (1981)
157. V.S. Stepanov: "Methods of Investigation of the Effectiveness of Energy Utilization in Technological Processes" (in Russian), Eleventh Int. Conf. Ind. Energy Econ., Berlin, Paper N A-1-9 (1984)
158. V.S. Stepanov: "Modeling and Study of Energy Use in Technological Processes at the Level of Their Ideal Analogs" (in Russian), Twelfth Int. Conf. Energetics, Pague, Paper N A-4/20 (1987)
159. V.S. Stepanov, T.B. Stepanova: "Analysis of Energy Use in Copper Production" (in Russian), Tsvetnye Metally, **53**(4), 79–83 (1978)
160. V.S. Stepanov, T.B. Stepanova: "Determination of the Energy Use Level in Metallurgical Productions Using Complete Energy Balance" (in Russian), Promyshlennaya Energetika, **44**(2), 8–10 (1987)
161. V.S. Stepanov, T.B. Stepanova: "Estimation of Costs of Energy Conservation Measures and Determination of Priority in Their Realization" (in Russian), Promyshlennaya Energetika, **44**(8), 7–10
162. V.S. Stepanov, T.B. Stepanova: "Estimation of Energy Conserved on the Basis of Modeling the Energy Consumption of Technological Processes in Ferrous Metallurgy" (in Russian), Promyshlennaya Energetika, **46**(4), 7–10 (1989)
163. V.S. Stepanov, T.B. Stepanova: "Energy Efficiency Increase of the Agglomeration Process in Lead Production" (in Russian), Promyshlennaya Energetika, **34**(12), 10–12 (1977)

164. V.S. Stepanov, T.B. Stepanova: "Investigation of Efficiency of Autogeneous Processes" (in Russian), Thirteenth Int. Conf. Ind. Energetics, Varna, Paper N A-2-28 (1990)
165. V.S. Stepanov, T.B. Stepanova: "Study of the Prospective Energy Consumption in Ferrous Metallurgy with Account of Technological Progress and Energy Conservation Policies in this Sector" (in Russian), Twelfth Int. Conf. Ind. Energetics, Prague (1987)
166. V.S. Stepanov, T.B. Stepanova: "Methods for Ascertaining the Potential for Energy Resource Conservation in Industry" (in Russian), in *Energy Systems: Control of Development and Operation* (Izd-vo SEI SO AN SSSR, Irkutsk 1986) pp. 65–74
167. V.S. Stepanov, T.B. Stepanova: "Technique for Distributing the Energy Costs in Complex Processing of Polymetallic Raw Material (in Russian), Izv. Vuzov, Energetika, **22**(12), 93–98 (1979)
168. V.S. Stepanov, T.B. Stepanova: *Potential and Actual Energy Conserved in Industry* (in Russian) (Nauka, Novosibirsk 1990)
169. S. Stepien: "Use of Steel and Iron Wastes in Ferrous Metallurgy" (in Polish), Gospodarka materialowa, **36**(6), 108–112 (1984)
170. A. Stodola: "The Cyclic Processes of the Gas Engine" (in German), Z. VDI, **32**(38), 1086–1091 (1988)
171. D.R. Stull, E.F. Westrum, G.C. Sinke: *The Chemical Thermodynamics of Organic Compounds* (J. Wiley, New York 1969)
172. M.W. Sussman: "Steady Flow Availability and the Standard Chemical Availability", Energy, **5**(8/9), 793–802 (1980)
173. J. Szargut: "Chemical Exergies of the Elements", Applied Energy, **32**, 269–286 (1989)
174. J. Szargut: "Exergy Balance of Metallurgical Processes" (in Polish), Arch. Hutn., **6**(1), 23–60 (1961)
175. J. Szargut: "Exergy Balance of Thermal Processes in Metallurgy" (in German), Freiberg. Forschungsh. B68, 81–103 (1962)
176. J. Szargut: "Exergy of Typical Raw Materials and Products in Metallurgy" (in German), Neue Hütte, **10**(5), 266–275 (1965)
177. J. Szargut: "International Progress in Second Law Analysis", Energy, **5**, 709–718 (1980)
178. J. Szargut: "Reference Level of Chemical Exergy", Arch. Termodynamiki, **9**(1–2), 41–52 (1988)
179. J. Szargut: *Heat Engineering in Metallurgy* (in Russian) (Metallurgiya, Moscow 1976)
180. J. Szargut: "The Potential Balance of Chemical Processes" (in Polish), Arch. Budowy Masz., **4**(4), 87–117 (1957)
181. J. Szargut: "The Potential Balance Resulting from the Second Law of Thermodynamics for Physical Processes" (in Polish), Arch. Budowy Masz. **3**(3), 231–276 (1956)
182. J. Szargut, C. Dziedziniewicz: "Available energy of Inorganic Chemical Substances" (in French), Entropie, No. 40, 14–23 (1971)
183. J. Szargut, D.R. Morris: "Calculation of the Standard Chemical Exergy of Some Elements and Their Compounds Based on Sea Water as the Reference Substance", Bull. Acad. Pol. Sci. Tech., **33**(5–6), 293–305 (1985)
184. J. Szargut, D.R. Morris, F.R. Steward: *Exergy Analysis of Thermal, Chemical and Metallurgical Processes* (Springer, Berlin, Heidelberg 1988)
185. Z. Szargut, S. Pasierb, J. Wandrasz: "Exergy Balance of Steel Plants" (in Polish), Hutnik, **31**(4), 123–128 (1964)
186. J. Szargut, R. Petela: *Exergy* (in Russian), (Energiya, Moscow 1968)
187. J. Szargut, T. Styrylska: "Approximate Evaluation of the Exergy of Fuels" (in German), Brennst. Wärme Kraft, **16**(12), 589–596 (1964)
188. *The Methodical Basis for Planning the Use of Secondary Energy Resources* (in Russian) (Energoatomizdat, Moscow 1987)

189. *Thermodynamic Properties of Substances*, Ed. V.P. Glushko, Vol. I–IV (in Russian) (Izd-vo AN SSSR, Moscow 1962)
190. *Thermodynamic Properties of Fuels and Burden Materials of Ferrous Metallurgy*: Reference Book (in Russian) (Metallurgiya, Moscow 1982)
191. Yu.A. Tikhomirov, A.V. Grachev, V.I. Pechnikov: "Energy Efficiency of Improving Steelmaking" (in Russian), Promyshlennaya Energetika, **42**(11), 2–3 (1985)
192. A.D. Tsoi, M.I. Tsysin, A.V. Rachinski, O.I. Golovanov: "Problems of Using Low-Potential Heat in the National Economy" (in Russian), Promyshlennaya Energetika, **41**(1), 9–12 (1984)
193. E.S. Ventsel: *Elements of Dynamic Programming* (in Russian) (Nauka, Moscow 1964)
194. N.M. Vilenski: *Rational Use of Secondary Energy Resources* (in Russian) (Metallurgyia, Moscow 1964)
195. H. Voigt: "Evaluation of Energy Processes Through Enthropy and Exergy", International Institute for Applied Systems Analysis Report No. RM-78-60 (November, 1978)
196. A.P. Voinov, L.I. Kuperman, S.P. Sushon: *Steam Boilers Fueled by Waste Gases* Manual. (in Russian) (Vysshaya Shkola, Kiev 1967)
197. M.P. Vukalovich: *Tables of Thermophysical Properties of Water and Steam* (in Russian) (Mashinostroyeniye, Moscow-Leningrad 1967)
198. M.P. Vukalovich, V.A. Kirillin, S.A. Remizov, V.S. Siletski, V.N. Timofeev: *Thermodynamic Properties of Gases* (in Russian) (Mashgiz, Moscow 1953)
199. M.P. Vukalovich, I.I. Novikov: *Thermodynamics* (in Russian) (Mashynostroyeniye, Moscow 1972)
200. D.J. Wilde: *Optimum Search Methods* (in Russian) (Nauka, Moscow 1967)
201. S. Yamauchi, K. Fueki: "New Thermodynamic Functions, Theta Function and Reference Exergy of Elements", Data for Science and Technology, Proc. of 7th CODATA Conf. (Kyoto, 1980) (Oxford University Press 1981) pp. 242–245
202. A.S. Yastrzhembski: *Technical Thermodynamics* (in Russian) (Gosenergoizdat, Moscow-Leningrad 1960)
203. M.Ya. Yudashkin: *Gas Cleaning in Metallurgy* (in Russian) (Metallurgiya, Moscow 1976)
204. O.V. Yuzov: *Efficiency of New Steelmaking Processes and Units* (in Russian) (Metallurgiya, Moscow 1977)
205. N.D. Zakharov: "Exergy of Fossil Fuels" (in Russian), Iz. Vuzov, Energetika, **14**(9), 63–67 (1970)

# Subject Index

MIX
Papier aus verantwortungsvollen Quellen
Paper from responsible sources
FSC® C105338

If you have any concerns about our products,
you can contact us on
**ProductSafety@springernature.com**

In case Publisher is established outside the EU,
the EU authorized representative is:
**Springer Nature Customer Service Center GmbH**
**Europaplatz 3, 69115 Heidelberg, Germany**

Printed by Libri Plureos GmbH
in Hamburg, Germany